ASCE Manuals and Reports on Engineering Practice No. 91

Design of Guyed Electrical Transmission Structures

Prepared by the
Subcommittee on Guyed Transmission Structures of the
Committee on Electrical Transmission Structures of
The Structural Engineering Institute of the
American Society of Civil Engineers

Published by

ASCE *American Society of Civil Engineers*

345 East 47th Street
New York, New York, 10017-2398

Abstract:

Guyed structures are commonly used to support electric transmission lines. They generally have the advantage of lightweight, erection ease, pre-assembly, and simple foundation design. There is a considerable range of applications, from simple guyed wood poles to the very large guyed steel latticed structures. This guide was prepared to supplement the various ASCE and IEEE guides for the design of electrical transmission structures. This publication describes the various types of guyed structures that have been used; presents typical guys and fittings; illustrates guy anchors and foundations; explores analysis and design techniques specific to guyed structures; discusses unique construction and maintenance problems; and displays both hand and computer calculations to illustrate some of the concepts discussed in the document.

Library of Congress Cataloging-in-Publication Data

American Society of Civil Engineers. Committee on Electrical Transmission Structures. Subcommittee on Guyed Transmission Structures.
Design of guyed electrical transmission structures / by the Subcommittee on Guyed Transmission Structures of the Committee on Electrical Transmission Structures of The Structural Engineering Institute of the American Society of Civil Engineers.
p.   cm. -- (ASCE manuals and reports on engineering practice ; no. 91)
ISBN 0-7844-0284-1
1. Electric lines--Poles and towers--Design and construction. 2. Guy anchors. I. Title. II. Series.
TK3242.A52 1997   97-26968
621.319'22--dc21   CIP

MANUALS AND REPORTS ON ENGINEERING PRACTICE

(As developed by the ASCE Technical Procedures Committee, July 1930, and revised March 1935, February 1962, April 1982)

A manual or report in this series consists of an orderly presentation of facts on a particular subject, supplemented by an analysis of limitations and applications of these facts. It contains information useful to the average engineer in his everyday work, rather than the findings that may be useful only occasionally or rarely. It is not in any sense a "standard," however; nor is it so elementary or so conclusive as to provide a "rule of thumb" for nonengineers.

Furthermore, material in this series, in distinction from a paper (which expresses only one person⌐s observations or opinions), is the work of a committee or group selected to assemble and express information on a specific topic. As often as practicable the committee is under the direction of one or more of the Technical Divisions and Councils, and the product evolved has been subjected to review by the Executive Committee of the Division or Council. As a step in the process of this review, proposed manuscripts are often brought before the members of the Technical Divisions and Councils for comment, which may serve as the basis for improvement. When published, each work shows the names of the committees by which it was compiled and indicates clearly the several processes through which it has passed in review, in order that its merit may be definitely understood.

In February 1962 (and revised in April, 1982) the Board of Direction voted to establish:

A series entitled 'Manuals and Reports on Engineering Practice,' to include the Manuals published and authorized to date, future Manuals of Professional Practice, and Reports on Engineering Practice. All such Manual or Report material of the Society would have been refereed in a manner approved by the Board Committee on Publications and would be bound, with applicable discussion, in books similar to past Manuals. Numbering would be consecutive and would be a continuation of present Manual numbers. In some cases of reports of joint committees, bypassing of Journal publications may be authorized.

#	MANUALS AND REPORTS OF ENGINEERING PRACTICE
10	Technical Procedures for City Surveys
13	Filtering Materials for Sewage Treatment Plants
14	Accommodation of Utility Plant Within the Rights-of-Way of Urban Streets and Highways
31	Design of Cylindrical Concrete Shell Roofs
33	Cost Control and Accounting for Civil Engineers
34	Definitions of Surveying and Associated Terms
35	A List of Translations of Foreign Literature on Hydraulics
36	Wastewater Treatment Plant Design
37	Design and Construction of Sanitary and Storm Sewers
40	Ground Water Management
41	Plastic Design in Steel-A Guide and Commentary
42	Design of Structures to Resist Nuclear Weapons Effects
45	Consulting Engineering-A Guide for the Engagement of Engineering Services
46	Report on Pipeline Location
47	Selected Abstracts on Structural Applications of Plastics
49	Urban Planning Guide
50	Planning and Design Guidelines for Small Craft Harbors
51	Survey of Current Structural Research
52	Guide for the Design of Steel Transmission Towers
53	Criteria for Maintenance of Multilane Highways
54	Sedimentation Engineering
55	Guide to Employment Conditions for Civil Engineers
57	Management, Operation and Maintenance of Irrigation and Drainage Systems
58	Structural Analysis and Design of Nuclear Plant Facilities
59	Computer Pricing Practices
60	Gravity Sanitary Sewer Design and Construction
62	Existing Sewer Evaluation and Rehabilitation
63	Structural Plastics Design Manual
64	Manual on Engineering Surveying
65	Construction Cost Control
66	Structural Plastics Selection Manual
67	Wind Tunnel Model Studies of Buildings and Structures
68	Aeration-A Wastewater Treatment Process
69	Sulfide in Wastewater Collection and Treatment Systems
70	Evapotranspiration and Irrigation Water Requirements
71	Agricultural Salinity Assessment and Management
72	Design of Steel Transmission Structures
73	Quality in the Constructed Project-a Guide for Owners, Designers, and Constructors
74	Guidelines for Electrical Transmission Line Structural Loading
75	Right-of-Way Surveying
76	Design of Municipal Wastewater Treatment Plants
77	Design and Construction of Urban Stormwater Management Systems
78	Structural Fire Protection
79	Steel Penstocks
80	Ship Channel Design
81	Guidelines for Cloud Seeding to Augment Precipitation
82	Odor Control in Wastewater Treatment Plants
83	Environmental Site Investigation
84	Mechanical Connections in Wood Structures
85	Quality of Ground Water
86	Operation and Maintenance of Ground Water Facilities
87	Urban Runoff Quality Manual
88	Management of Water Treatment Plant Residuals
89	Pipeline Crossings
90	Guide to Structural Optimization
91	Design of Guyed Electrical Transmission Structures
92	Manhole Inspection and Rehabilitation

CONTENTS

Preface ix

Chapter
  1 Introduction 1

  2 Guyed Structures Configurations 3
    2.1 General 3
    2.2 Single Poles or Masts 4
        2.2.1 Guying configurations 4
        2.2.2 Pole or mast base 5
        2.2.3 Limits of use 6
    2.3 Stub Poles 6
    2.4 H-Frames (Multi-Pole Structures) 7
    2.5 Rigid Frames 8
        2.5.1 Guyed rigid latticed portal 8
        2.5.2 Guyed rigid Y 9
        2.5.3 Guyed delta 9
    2.6 Masted Towers 9
        2.6.1 Guyed portal 11
        2.6.2 Guyed V 11
        2.6.3 Cross rope 13
        2.6.4 Guyed hinged Y 13

  3 Guys and Guy Fittings 15
    3.1 Guy Materials 15
    3.2 Guy Fittings 17
    3.3 Tensioning Devices 20

  4 Guy Anchors and Foundations 21
    4.1 Deadman Anchors 21
    4.2 Screw Anchors 22
    4.3 Grouted Anchors 23

  5 Analysis 27
    5.1 Cable Behavior 27
    5.2 Poles or Latticed Masts with Single Guy Attachment Point 31
        5.2.1 Single guy level and hinged base 31
        5.2.2 Single guy level and fixed base 32
            5.2.2.1 Effect of preload or foundation movement 35
            5.2.2.2 Effect of temperature 36

5.3 Poles or Latticed Masts with Multiple Guy
Attachment Points 36
   5.3.1 Multi-guy levels and hinged base 36
   5.3.2 Multi-guy levels and fixed base 37
5.4 Structures with Four Guys 37
5.5 Buckling Strength of Poles and Latticed Masts 38
   5.5.1 Pole buckling strength 38
   5.5.2 Equivalent beam model for latticed masts 40
5.6 Computer Modeling 42
   5.6.1 Linear computer analysis 42
   5.6.2 Nonlinear computer analysis 43
   5.6.3 Modeling hints and details 44
      5.6.3.1 Special considerations for guyed
concrete poles 45

6 Design 47
6.1 Loadings and Design Restrictions 47
6.2 Guy Clearances 48
   6.2.1 Electrical clearances 48
   6.2.2 Mechanical clearances 48
6.3 Guy Design 49
   6.3.1 Guy pretension 49
   6.3.2 Allowable tensions 50
   6.3.3 Guy slope 50
6.4 Guy Anchorage 51
6.5 Connections 52
   6.5.1 Guy connections to poles 52
   6.5.2 Connections in latticed structures 52
6.6 Structural Design of Poles and H-Frames 53
6.7 Structural Design of Latticed Rigid Frames and
Masted Towers 54

7 Construction and Maintenance 55
7.1 Design Considerations 55
7.2 Construction Considerations 55
7.3 Guyed Poles 56
   7.3.1 Erection methods 56
   7.3.2 Guy installation 56
7.4 Guyed Rigid Frames and Masted Towers 57
   7.4.1 Erection methods 57
   7.4.2 Crane erection 57
   7.4.3 Helicopter erection 58
   7.4.4 Guy installation 58
      7.4.4.1 Traditional method 59
      7.4.4.2 Alternate method 59
      7.4.4.3 Guy pretension 60

CONTENTS vii

7.5 Erection Tolerances 60
   7.5.1 Guyed poles and H-frames 60
   7.5.2 Guyed rigid frames and masted towers 60
7.6 Inspection and Maintenance 61

8 Examples 63
8.1 Wood Poles 63
   8.1.1 Dead-end pole with in-line guys in single
   vertical plane 63
      8.1.1.1 Analysis and buckling capacity by
      manual methods 64
      8.1.1.2 Analysis and buckling capacity by
      nonlinear computer analysis 65
   8.1.2 Ninety-degree angle pole with in-line guys 66
      8.1.2.1 Analysis and buckling capacity by
      manual methods 66
      8.1.2.2 Analysis and buckling capacity by
      nonlinear computer analysis 66
8.2 Tubular Steel Poles 67
   8.2.1 Bisector guyed pole 68
   8.2.2 Effect of guy properties on behavior of pole 70
8.3 Guyed V 72
   8.3.1 Analysis for high wind loads 72
   8.3.2 Analysis for unbalanced longitudinal load on
   outer phase 73
   8.3.3 Analysis for combination of vertical, transverse,
   and longitudinal loads 73
   8.3.4 Design of mast for guyed V 74
8.4 Guyed Delta 76

Appendices
   A **REFERENCES** 79

   B **NOTATION** 81

Index 83

PREFACE

In 1991 the ASCE Committee on Electrical Transmission Structures (CETS) recommended that a subcommittee be formed to prepare a guide for the design of guyed transmission structures. The CETS has been or is being involved with the publication of design guides for the structural and geometric design of transmission or substation structures. Guides are or will soon be available for the design of steel transmission towers, tubular steel poles and frames, prestressed concrete poles and frames, as well as substation structures. It was felt that none of the existing or upcoming guides contained enough information regarding guyed structures. Therefore the CETS Subcommittee on Guyed Transmission Structures was established in 1991 to prepare this publication.

This guide represents the consensus of opinion of the subcommittee and although the subject matter of the guide has been thoroughly researched, its application should come only after sound engineering judgment has been used.

The many and unique contributions of H. Brian White to this document through his work with the subcommittee and his paper on guyed structures (White 1993) are greatly appreciated.

The subcommittee wishes to thank the Peer Review Committee for its contributions to the final draft of this document: Leon Kempner (Chair), Lindsay Esterhuizen, Jake Kramer, and Goetz Schildt. It also wishes to acknowledge the assistance of the three chairmen of the CETS during whose tenure this guide was conceived and developed: Anthony DiGioia, Alain Peyrot, and Leon Kempner.

Respectfully submitted:

Subcommittee on Guyed Transmission Structures
Committee on Electrical Transmission Structures

Clayton L. Clem	Michael D. Miller
James S. Cohen	Robert E. Nickerson (Vice Chair)
Martin L. De La Rosa	Alain H. Peyrot (Chair)
Michael Gall	Ronald E. Randle
Magdi F. Ishac	Randall L. Samson
Massoud Khavari	Joe Springer
Richard Kravitz	Larry D. Vandergriend
Jerry Lembke	H. Brian White
Robert M. McCafferty	Jerry Wong

Chapter 1

INTRODUCTION

Guyed structures are commonly used to support electric transmission lines. They generally have the advantage of light weight, erection ease, pre-assembly, and simple foundation design. There is a considerable range of applications, from the simple guyed wood poles to the very large guyed steel latticed structures. Although the advantages of guying a simple pole structure are well understood by designers, there is generally a poor understanding of the potential benefits or problems associated with other types of guyed structures. In addition, there is currently very little published information describing appropriate analysis or design techniques applicable to all guyed transmission structures.

This manual was prepared to supplement the various ASCE and IEEE guides for the design of electrical transmission structures (ANSI / ASCE 1991; ASCE 1990; ASCE 1991; IEEE 1985; IEEE 1991) with much needed information on the proper use and design of guyed structures. Guyed structures made of all common materials are covered (reinforced or prestressed concrete, aluminum, steel, or wood).

Section 2 describes the various types of guyed structures that have been used and their relative advantages and disadvantages. Typical guys and fittings are presented in Section 3 and guy anchors and foundations are illustrated in Section 4. In Sections 5 and 6, analysis and design techniques specific to guyed structures are presented. Section 7 discusses unique construction and maintenance problems presented by guyed structures. Finally, in Section 8, both hand and computer calculations illustrate some of the concepts discussed in the document.

Chapter 2

GUYED STRUCTURES CONFIGURATIONS

2.1 GENERAL

The overall configuration of a guyed structure is a function of line voltage, electric air gap clearance requirements, ground clearance requirements, electric and magnetic field limits, insulation requirements, structural loading, number of circuits to be supported, conductor phase arrangement, configuration of the circuits, right-of-way requirements, and aesthetic design criteria.

Guyed or unguyed, transmission structures can be classified as suspension, strain, or dead-end. The conductor phases pass through, or by, a suspension structure and are suspended by insulators. There is generally no longitudinal load from imbalance of cable tensions on a suspension structure as long as the conductor system is intact, adjacent towers have not fallen, or no unbalanced ice condition exists. In a strain structure, the conductors are directly attached to the structure through in-line insulators. Thus any imbalance of cable tension has to be carried by the structure. A dead-end structure is similar to a strain structure except that its design includes conditions with the conductors and ground wires pulling on only one side of the structure. In strain and dead-end guyed structures with permanent unbalanced tensions and in all angle structures, there is permanent loading in the guy system requiring careful consideration of how potential creep of the anchors can change the distribution of forces.

The advantages of guyed structures are many. When conditions of use are suitable, large economies can result if the best type is selected to suit the many and varying conditions that can face the line designer. Terrain type (flat, rolling, mountainous), access and transport situation, erection techniques, procurement and easements, structural loadings, and electrical constraints are all parameters that can influence the selection of the structure type.

A guyed structure is not always the best solution as restrictions on the use of guys are sometimes sufficiently limiting to lead to the selection of other types of support. For example, guyed structures will almost always require a larger site area for the anchoring of the guys and this alone may rule out their use in farmlands where large equipment is used. The guys can present a hazard to the equipment and, conversely, large machinery may pose a threat to the guys. However, this may not be decisive; witness the 25-year old 765 kV guyed V tower lines successfully running through Indiana and Ohio farmland (Kravitz and Samuelson 1969). The guys of these towers were steeper than optimal to better accommodate farm equipment. In urban and suburban areas, the right to get passage for a new line may depend on minimizing the occupied land area and this may force the use of compact rigid self-supporting construction.

It is convenient to classify guyed structures into general categories: (1) single poles or masts, (2) stub poles, (3) H-frames, (4) guyed rigid frames, and (5) masted towers. There are similarities in the way the single poles, stub poles, and H-frames are analyzed or designed as they all use straight poles as their main components. Rigid frames (as defined herein) and masted towers usually do not use poles.

2.2 SINGLE POLES OR MASTS

Depending upon strength, material, and height limitations, a single guyed pole or mast can be designed to support one overhead ground wire and one or more phases or circuits in a vertical profile. Guyed poles are most common at line angles, strain, or dead-end locations where large permanent loads from cable tensions are present. Angle poles can be used either in a running angle (suspension) condition or in a strain condition. Guyed single poles of wood, concrete, or tubular steel are the most common types of guyed structures. Latticed masts can be used instead of poles.

2.2.1 Guying Configurations

Guying on a running angle structure (Fig. 2-1) is generally set against the resultant transverse loading in the direction opposite to the bisector of the line angle. For strain structures, true in-line (head and back) guys, modified in-line guys, bisector or bisector with in-line guys can be used (Fig. 2-2). The addition of a bisector guy is generally used for small line angles (say, less than 30 degrees) because of the lack of lateral stiffness of the in-line guys. The configuration of Fig. 2-2(c) provides little longitudinal capacity in case of broken conductors or unbalanced ice events. Guy slopes, orientations, and sizes affect the guys' effectiveness and consequently the distribution of forces between the guys and the structure as discussed in Section 5.

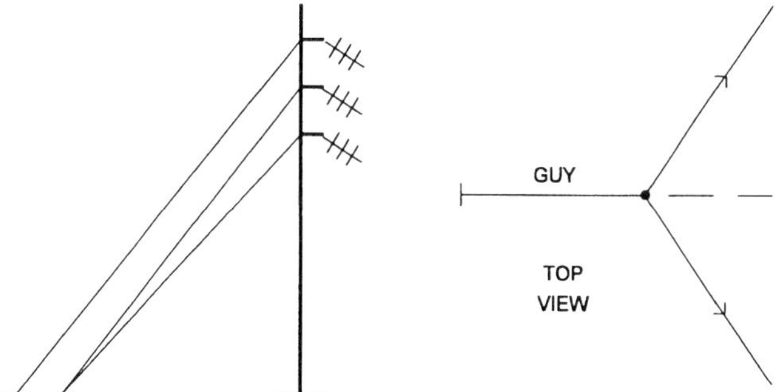

FIG. 2-1. Bisector Guyed Running Angle Pole.

Guying can be installed at the ground-wire level and each conductor phase level to minimize bending in the pole, or at intermediate positions. This is generally a function of allowable structure deflection and material strength.

2.2.2 Pole or Mast Base

Guyed single poles or masts can have pinned, fixed, or partially fixed bases. There is an axial compression load and a shear force at a pinned base. In addition, there is a base moment for partially or fully fixed bases. The force distribution between guys and pole and the ultimate buckling capacity of the pole are affected by the base condi-

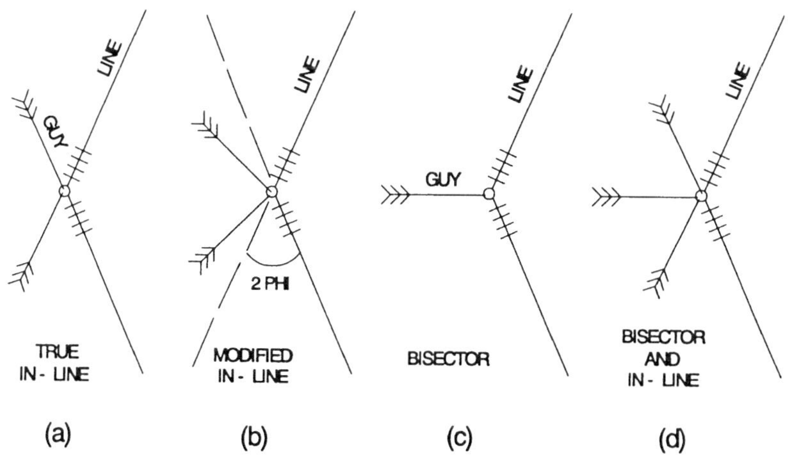

FIG. 2-2. Guy Arrangements at Small Line Angle.

tion, which is therefore an important parameter of a guyed pole analysis (see Section 5).

Fixed or partially fixed bases can be furnished principally by either method (1) direct embedment of the pole with sufficient depth to develop the groundline reactions, or (2) rigidly connecting the pole to a moment-resisting foundation.

2.2.3 Limits of Use

Naturally grown round wood poles are generally restricted by strength and height. Engineered wood poles such as laminated wood poles can be manufactured for higher strength and heights up to 36 m (120 ft). However, they are typically manufactured to round wood pole equivalent sizes. Longer lengths are possible with manufactured wood poles, but shipping and handling may require splices. Guyed wood poles are most commonly used at 115 kV and below, but have also been used at voltages up to 345 kV.

Due to initial cost, guyed tubular steel poles are generally not used in low voltage lines (69 kV and below). However, if maintenance, life expectancy, and reliability are considered, they can be competitive with other materials. Tubular steel structures are used in transmission lines through 500 kV.

Concrete poles are available in two basic types: static cast square poles and spun cast round poles. Both types are generally prestressed and produced in one piece. Single-piece poles are generally available in lengths up to 36 m (120 ft). Static cast are typically heavier than spun cast, especially as height and class increase; however, static cast generally have lower cost. Due to weight, concrete poles may not economically be shipped long distances or used in difficult areas. Therefore the selection of pole type depends essentially upon what is economically available near the job site. Special measures to prevent damage or breakage of concrete poles during transportation may be required. Since the major load in a guyed pole is axial, their compressive strength makes concrete poles an excellent choice for guyed construction. Although some concrete has been used in Extra High Voltage (EHV) lines, most use is in lines of up to 345 kV.

From a structural standpoint, any reasonable configuration of height and column width can be used for a guyed latticed mast.

2.3 STUB POLES

Guy stub poles are used as part of the guying system of another structure. They are used where clearance problems require that the slope of the guy changes as shown in Fig. 2-3. The stub pole may be directly embedded or supported on a foundation, and either self-supporting or guyed. The stub pole should be considered a compo-

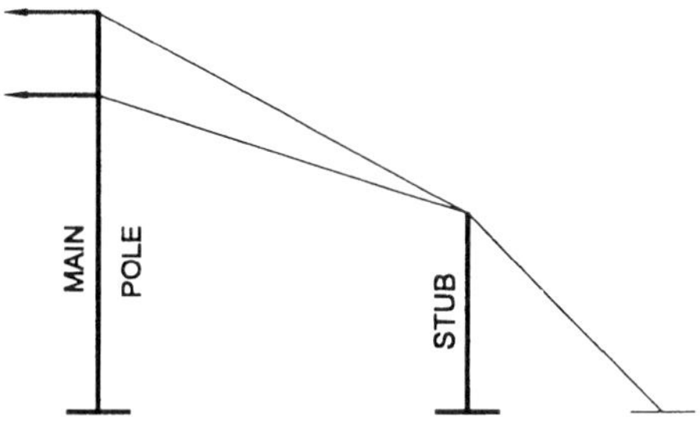

FIG. 2-3. *Stub Pole.*

nent of the overall guyed system which includes the main structure, the span guy, the guy stub pole, and the guy of the guy stub (if used).

2.4 H-FRAMES (MULTI-POLE STRUCTURES)

H-frames are defined in this document as structures made up of two or more poles (wood, steel, or concrete) connected with one or more horizontal crossarms and possibly X-braces. Although they most commonly support electrical circuits in a flat configuration, delta configurations are possible. H-frames are most often guyed to reduce uplift, bending, deflection, or the likelihood of aeolian vibration of the poles. H-frames can also have cross guys from pole to pole to reduce the effects of the guys on the poles and cross braces. H-frames are sometimes guyed longitudinally (along the line) at each structure or at intervals to prevent long cascading failures.

Guys are generally placed externally on the poles at the connection of the X-braces to reduce bending moments in the poles (Fig. 2-4). H-frames can also be guyed internally to replace the normal

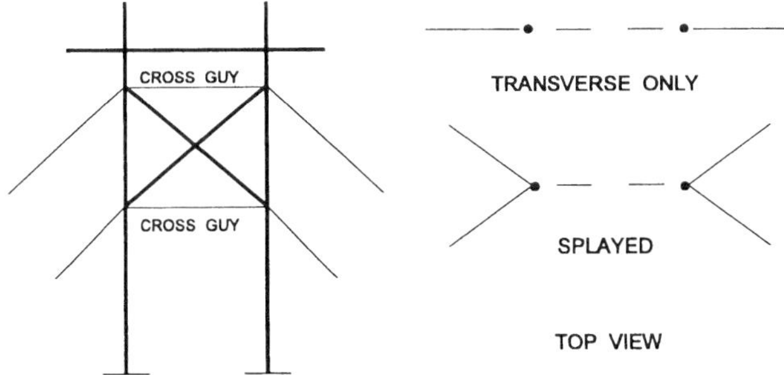

FIG. 2-4. *Guyed H-Frame.*

X-brace. However, if this is done, then one pole carries the majority of the load when one guy goes slack under the applied loading. As with single guyed poles, the bases of the poles in an H-frame can be pinned, fixed, or partially fixed.

Guyed wood H-frames have been used extensively from 69 kV through 345 kV and are limited by the height and strength of the poles that can be obtained. These structures are generally restricted for use as tangent suspension structures and small to medium line angle suspension structures. Guyed concrete or steel H-frames can be designed for longer spans than wood H-frames. However, these longer spans generally require larger rights-of-way and greater guy loads. Hillside conditions should be investigated as the stress distribution for unequal height poles can be substantially different than for equal height poles.

2.5 RIGID FRAMES

This category includes steel latticed or tubular rigid frames (excluding multi-pole structures) stabilized with at least four guys splayed in the transverse and longitudinal directions to give support in all directions. The four-guy arrangement also provides torsional strength and rigidity when properly detailed.

2.5.1 Guyed Rigid Latticed Portal

The guyed rigid latticed portal frame is similar to an H-frame except that poles and crossarms are made up of a rigidly connected assembly of latticed steel angle member sections. This is shown in Fig. 2-5. It can be guyed internally or externally.

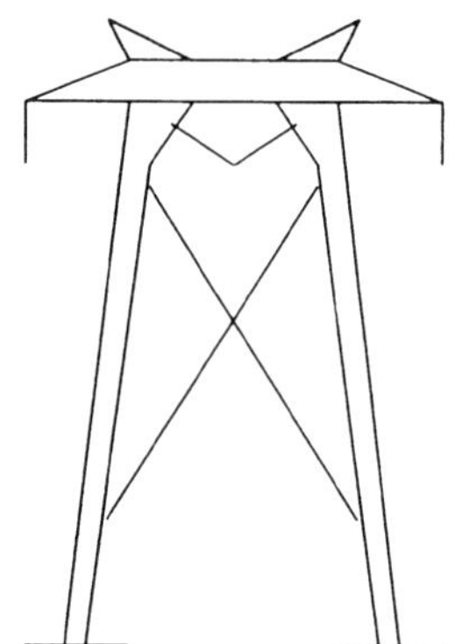

FIG. 2-5. Internally Guyed.

2.5.2 Guyed Rigid Y

Examples of outlines of guyed rigid Ys are shown in Fig. 2-6, where the Y and the crossarm can be latticed or tubular steel. The stem and upper branches of the Y are rigidly connected.

2.5.3 Guyed Delta

The single-circuit guyed latticed delta is derived from the single-circuit latticed delta tower by replacing its body by a mast tapered towards the ground and four guys, as shown in Fig. 2-7(a). Clearance to guys and torsional stiffness are improved by crossing the guys. Torsional resistance to a single longitudinal load on an outer phase may suffer as rotation will cause the lines of action of the resisting guys to get closer together with significant second order increases in guy and anchor loads. This is one of the guyed latticed towers that clearly justifies a geometrically nonlinear elastic computer analysis. Some designers use secondary stabilizing arms as shown in Fig. 2-7(b) to prevent this secondary increase in loads and distortion as the arms act to maintain guy separation with twist. The guyed delta can also be fabricated with tubular steel as shown in Fig. 2-8.

2.6 MASTED TOWERS

This category includes structures made up of pin-connected straight latticed masts stabilized by guy cables. Unlike the structures in the previous categories, masted towers are near statically determinate; that is, the axial compression forces in the masts and the tensions in the cables can be calculated manually with good accuracy using principles of statics. Normal guy strains and structure distortions have negligible effect on overall strength.

The first three structures in this category, the guyed portal, the guyed V, and the Cross Rope Suspension (CRS) or ''Chainette'' are each based on the concept of two tripods, each tripod consisting of a

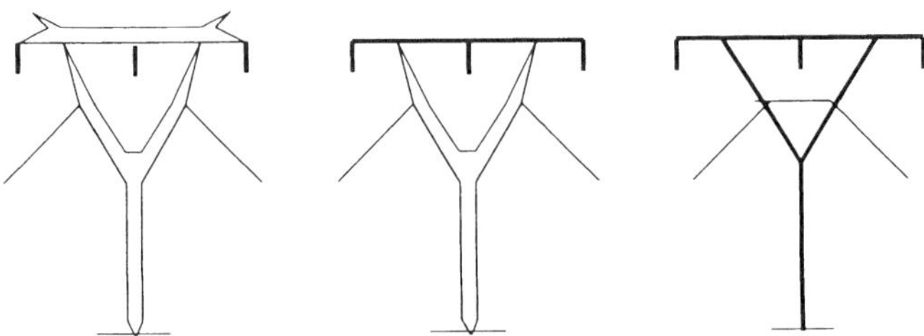

FIG. 2-6. Guyed Rigid Ys.

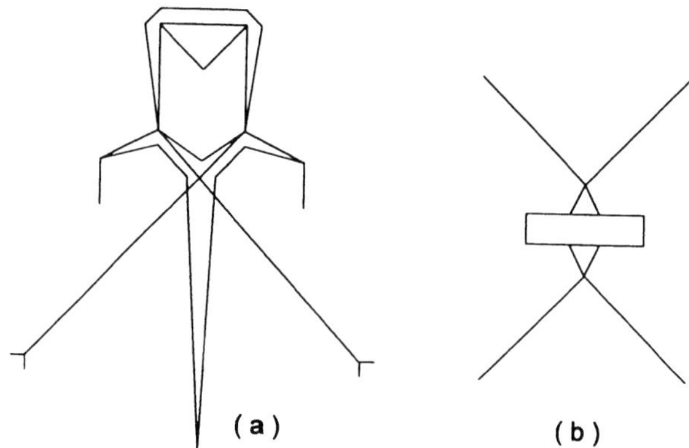

FIG. 2-7. *Latticed Guyed Delta.*

compression mast and two guys. The crossarm of the portal holds the mast tops apart; the guyed V crossarm holds the tops of the masts together. For the CRS, the mast tops are held against the outward pull of the guys by the inward cross-rope pull which results from the vertical loads of the insulators. All three structures will be stable in the event of removal of one guy as the conductors and ground wires (if used) will help to stabilize the structure although with a reduced transverse capacity.

The guyed mast family of towers is the most common type in the world for EHV lines (White 1993). The Guyed Portal has been used extensively since the 1930s in Northern Europe, Russia, and neighboring countries, and the Guyed V extensively elsewhere. The CRS

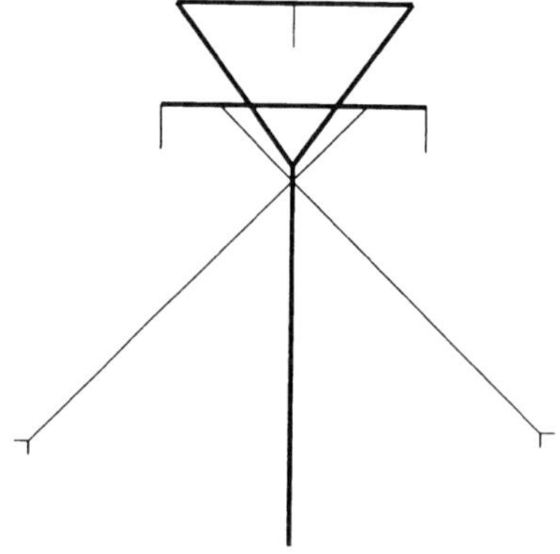

FIG. 2-8. *Tubular Delta.*

has recently been used at EHVs where the simple cross-rope wire assembly readily replaces the otherwise massive crossarms.

2.6.1 Guyed Portal

The guyed portal (Fig. 2-9) fits best on flat ground, and if it is not too high, the four guys can be brought to two guy anchors (Fig. 2-9(b)) for a saving in cost and land use. However, at a given voltage and as the height increases, the transverse spacing between each mast footing and the guy anchors does not increase proportionally with excessively steep guys for the higher towers in the set. Guys that are too steep cause high loads in the masts, guys, and anchors. The height limit can be overcome by crossing the guys and using four separate anchors.

Use on transversely sloping terrain poses the biggest problems for the portal arrangement for with the unequal length masts, ground assembly and rotation up into place becomes very difficult. This structure is also sensitive to torsional loads as the lines of action of the resisting guys come closer together as twisting increases.

2.6.2 Guyed V

The guyed V (Fig. 2-10) is a derivative of the guyed portal wherein the bases of the mast are brought to a near single point but the guys are splayed outwardly. This removes the problem of the guyed portal unequal mast length during construction in sloping terrain. They can also be used in rough terrain when the guy slopes are steepened sufficiently so that the guy leads do not become excessively long. However, steepening of the guys will result in higher masts, guys, and anchor loads. Guyed V structures are not very sensitive to normal anchor to base differentials.

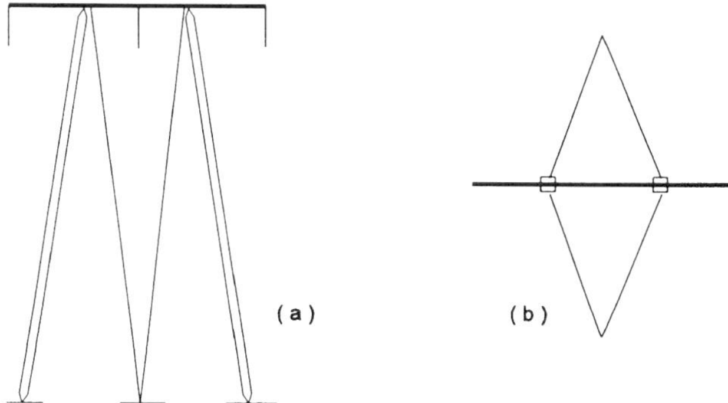

FIG. 2-9. *Internally Guyed Portal.*

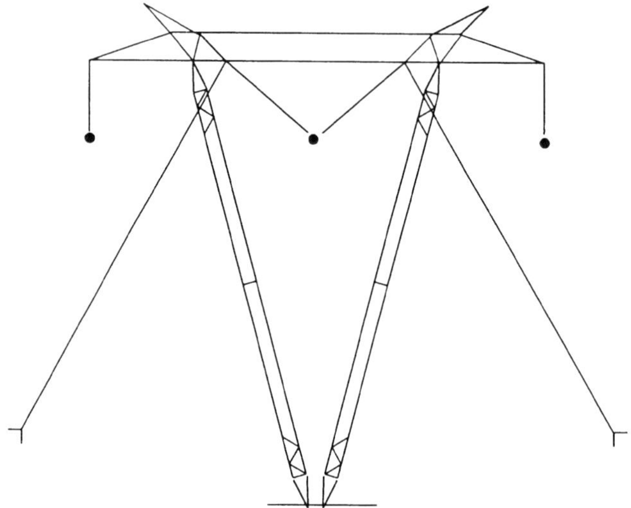

FIG. 2-10. *Guyed V.*

In order to minimize phase spacing at high voltage (765 kV), a guyed V with haunches (Fig. 2-11) was developed (Kravitz and Samuelson 1969) whereby the masts are terminated, not at the crossarm level, but at the base of the haunches. This works to keep the phase spacing to a minimum, however, at the possible sacrifice of additional structure weight in the crossarm and possible additional longitudinal bending moments in the masts.

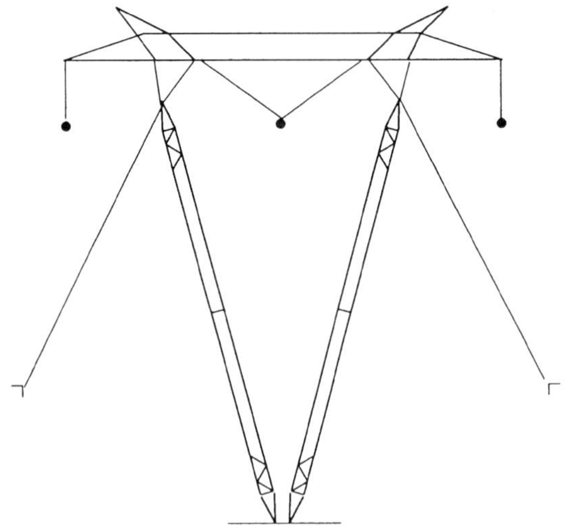

FIG. 2-11. *Guyed V with Haunches.*

2.6.3 Cross Rope

The CRS configuration is simple in both concept and design (Fig. 2-12). A small diameter spacer or construction cable is used between the masts for aligning and tensioning the guy system before the weight of the conductor system is in place and able to provide a balancing tension. It also maintains a constant distance between the mast tops so that the pre-cut and pre-assembled cross-rope assembly can be fit accurately. The CRS is probably the lightest, lowest cost, and easiest to erect structure for EHV lines. Its strength can match any rational load specifics and it automatically develops great failure containment properties. The CRS requires a large spacing of anchors at each structure location but the possibility of reduced phase spacings, an important parameter regarding Surge Impedance Loading (SIL), may also permit a reduction of right-of-way requirements between structure positions (White 1993).

The original CRS with a six-part rope suspension as shown in Fig. 2-12 is suitable for galloping conditions. A simple and less costly single rope suspension has been used with insulator assemblies attached to the cross rope with inverted suspension clamps (Behncke et al. 1994). A delta conductor configuration is also possible.

2.6.4 Guyed Hinged Y

The guyed hinged Y structure can be configured in two basic ways as shown in Fig. 2-13(a) and (b). A guyed hinged Y is essentially a small guyed V mounted (hinged) on top of a guyed vertical stem. A

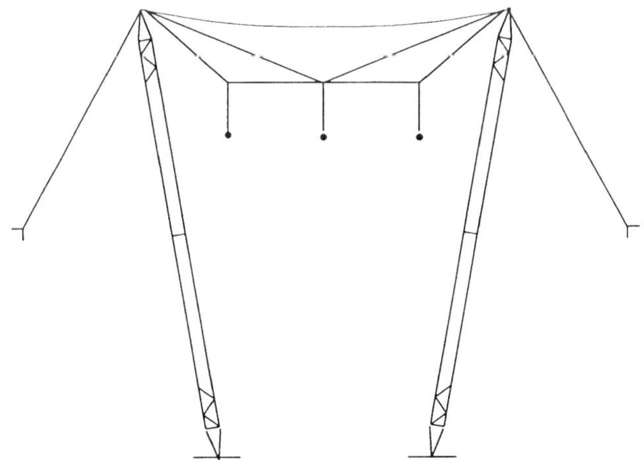

FIG. 2-12. Cross-Rope Suspension Tower.

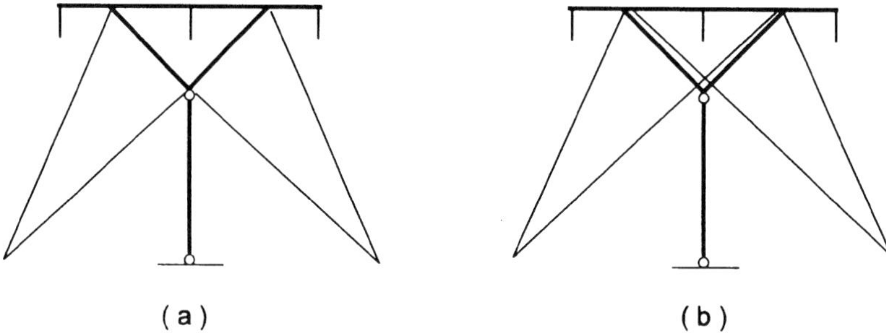

FIG. 2-13. *Guyed Hinged Y.*

minimum of eight (8) guys is required although connecting to only four anchors. The structure of Fig. 2-13(a) will collapse should one of the lower guys be removed. The Y of Fig. 2-13(b) will not fall with the removal of two of the eight guys because the connection at the base of the V is a hinge and not a pin. The hinged connection can transfer longitudinal moments from the V to the stem.

Chapter 3

GUYS AND GUY FITTINGS

The effective guy modulus of elasticity needed for analysis is defined in Section 5.1. Guy pretension and maximum working tensions are discussed in Sections 6.3.1 and 6.3.2. Guy installation is discussed in Sections 7.3.2 and 7.4.4.

3.1 GUY MATERIALS

Guys are typically of strand construction which permits sufficient flexibility for installation, the greater individual wire sizes giving better corrosion resistance and higher stress modulus of the complete stranding than do wire ropes. Typical strandings are shown in Fig. 3-1. Wire ropes are made of multiple strands but are not normally used in transmission structures. These ropes are more flexible as they are made of smaller individual wires. They are also more susceptible to corrosion and the effective modulus of elasticity of the rope is lowered in an almost direct relationship to the flexibility.

Galvanized high-strength strandings are most common, although some use is made of Alumoweld strandings where corrosion conditions are thought to be severe. A 7-wire stranding is practical to about 12 mm overall but beyond that, the individual wire size exceeds 4 mm and the stranding becomes very stiff and hard to handle, thus the change to 19 strands and subsequently to 37.

Some typical EHS (Extra High Strength) and some small Canadian Grade 1300 strand sizes (CAN/CSA-G12–92; ASTM A475–89; ASTM A586–92) and Rated Breaking Strengths (RBS) are shown in Tables 3-1 and 3-2, the sizes up to 12 mm and possibly 15 mm being used for low voltage construction, whereas the sizes of 19, 25, 28, 32, and even up to 35 mm are usually found on guyed EHV structures. The values shown are for information only and may not be applicable to designs.

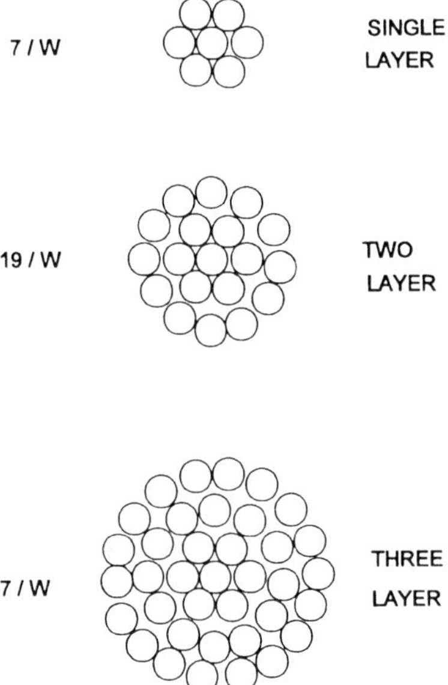

FIG. 3-1. Strand Configuration.

TABLE 3-1 Typical Guy Wires—EHS Grade
        (Imperial Sizes)

Diam. (mm)	Diam. (in.)	No.	Rated Breaking Strength (kN)	Rated Breaking Strength (kips)
6.35	1/4	3	30.0	6.7
7.94	5/16	3	40.4	9.1
9.52	3/8	3	52.4	11.8
4.76	3/16	7	17.7	4.0
6.35	1/4	7	30.0	6.7
7.94	5/16	7	49.8	11.2
9.52	3/8	7	68.5	15.4
12.7	1/2	7	119.6	26.9
15.88	5/8	7	188.6	42.4
14.29	9/16	19	149.9	33.7
19.05	3/4	19	259.3	58.3
25.40	1	19	464.8	104.5
28.58	1 1/8	37	581.7	130.8
31.75	1 1/4	37	721.4	162.2

TABLE 3-2 Typical Guy Wires—Grade 1300
(CSA-G12 Metric Sizes)

Nominal Diam. (mm)	Actual Diam. (mm)	No. Strands	Rated Breaking Strength (kN)
6	6.3	7	30
7	7.2	7	39
8	8.49	7	53
9	9.0	7	61
10	10.8	7	88
12	12.6	7	120

3.2 GUY FITTINGS

There are many different and suitable termination devices (see Fig. 3-2) for the smaller and relatively flexible strand sizes up to about 9 or even 12 mm diameter. Bolted clamps, fittings with wedges, preformed grips, and countless other types have been used with success and the selection depends on local availability and confidence in long-term holding strength.

The overall rigidity of strandings much above 12 mm diameter precludes devices such as the three-bolt clamp that must deform the strand to perform. Clamps that depend on surface friction can become very lengthy on large strandings as the needed holding strength varies as the diameter squared whereas the surface area varies only with the diameter.

Standard preformed grips or helical wire terminals perform well up to about 25 mm stranding diameter. Beyond that point, special long grips are required or the twisted grip part may have to be supplemented by a cone wedge device. Preformed grips may be reapplied up to two times for the purpose of retensioning with no significant loss of holding strength, but if installed for more than about three months, they should be replaced after removal. The apparent and actual ease of installation and removal sometimes attracts vandalism which can be thwarted by compressing a ring clip to the upper end of the assembly.

Swaged or compressed fittings are economical and easily applied in the shop or field but care must be taken in the selection of the steel. Some compressed fittings have cracked in brittle fracture under a combination of extremely low temperature and high internal stress levels that are produced during the swaging operation. This problem is no different than that of correct steel selection for the dead-end fittings for the conductor although larger stronger fittings are sometimes required for the guys.

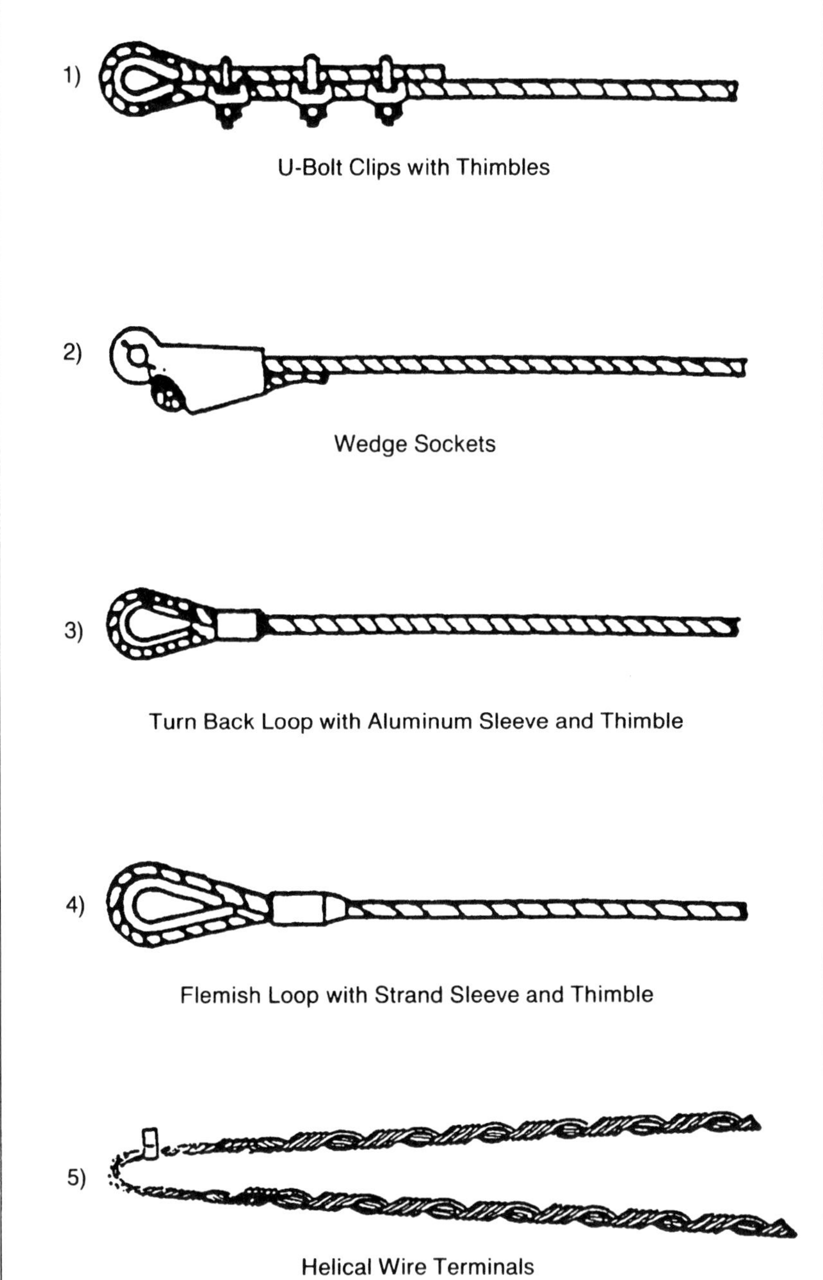

FIG. 3-2. *Typical Guy Fittings (1 of 2).*

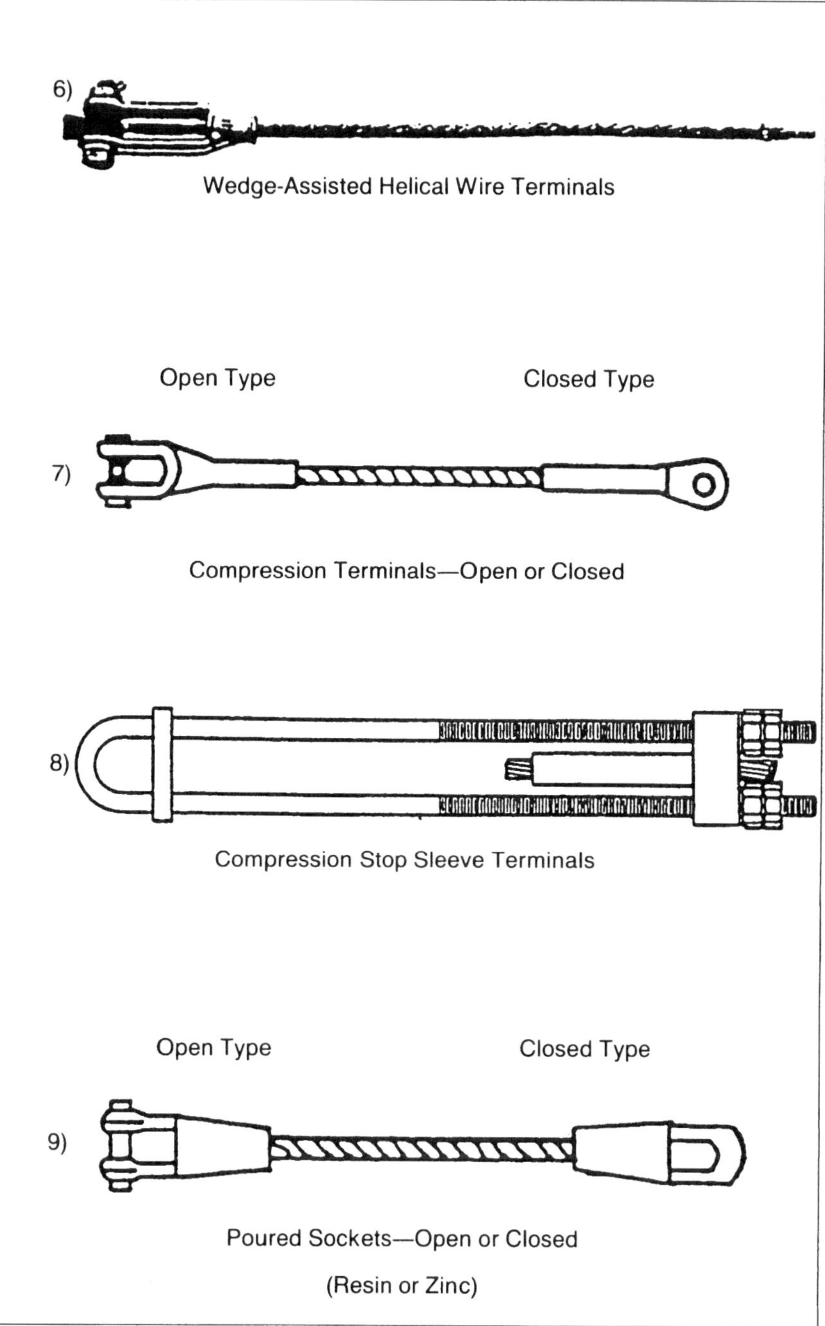

FIG. 3-2. *Typical Guy Fittings (2 of 2).*

Poured zinc sockets are sometimes used on the larger strandings where preformed grips do not appear adequate. They require considerable skill and care during fabrication, for the wire strands must be cleaned, individually splayed, and the entire assembly heated to ensure that the molten zinc flows down into the interstices of the wires. Voids will reduce holding power and may result in premature corrosion which will be hidden and difficult to detect. Failures have occurred due to poor quality control during the field installation of these devices.

3.3 TENSIONING DEVICES

Guy installation and tensioning are discussed in Sections 7.3.2 and 7.4.4. Permanent tensioning devices are usually installed only at the lower ends of the guys.

Guyed structures have traditionally been erected with guys attached at their tops and with temporary fittings and tackle to the anchors while the structure is made almost plumb. Permanent grips are installed and final positioning and pretensioning done with some form of threaded device. These can be simple threaded extensions of anchor rods, turnbuckles which can be very expensive in the larger sizes, or threaded U-bolts with keepers.

Chapter 4

GUY ANCHORS AND FOUNDATIONS

This section describes typical types of guy anchors that are commonly used today. It is not the intention of this guide to provide direction for the design of the anchor or foundation since there are other applicable guidelines presently available such as the *IEEE Trial-Use Guide for Transmission Structure Foundation Design* (IEEE 1985).

Guy anchors are foundations used to resist the tensile force imposed by the guy load. The load will result in both an uplift and horizontal force on the guy anchor. There is a variety of anchor types available to resist this load. The type of anchor selected is dependent on guy load, guy angle, soil conditions (peak and residual shear strength), and topography. Typical anchorages include grouted soil anchors, grouted rock anchors, spread steel or concrete plate anchors, screw anchors, concrete deadman and prestressed anchor blocks, and so on.

The design of an efficient guy anchor requires information on the shear and tensile strength of the soil or rock. Frequently this information is not available and thus it is difficult to determine the theoretical design capacities. Guy anchors can be proof tested to verify capacity in doubtful conditions and to "set" the anchor after installation. Such a test procedure is possible with anchors as permanent movements do not affect subsequent performance. Prior to installation of guy anchors, setting tolerances should be determined, a subject discussed in Section 7.5.

4.1 DEADMAN ANCHORS

Deadman anchors (also referred to as log anchors) are usually constructed by excavating a trench or hole into which an anchor is inserted and backfilled with compacted soil or concrete. The capacity of these types of anchors is based upon their uplift limitations and lateral shear strength of the soil. The capacity is also dependent on the degree and adequacy of compaction of the backfill. Figure 4-1 shows a number of various deadman-type anchors. Such anchors can be prestressed to limit uplift deformation. Log anchors require under-excavation with hand work and may be restricted by safety considerations. Plate anchors that require separate excavations for the anchor and the anchor rod (not illustrated) are described in the IEEE Guide (IEEE 1985). These plate anchors differ from spread anchors in that the in situ soil properties are used in calculating the capacity of the anchor.

4.2 SCREW ANCHORS

Screw anchors consist of a steel shaft fitted with one or more helixes. Typically this type of anchor is installed using power digging equipment such as an auger truck. This allows the application of both torque and axial load while pushing and rotating the shaft. This combined operation minimizes the disturbance of the surrounding

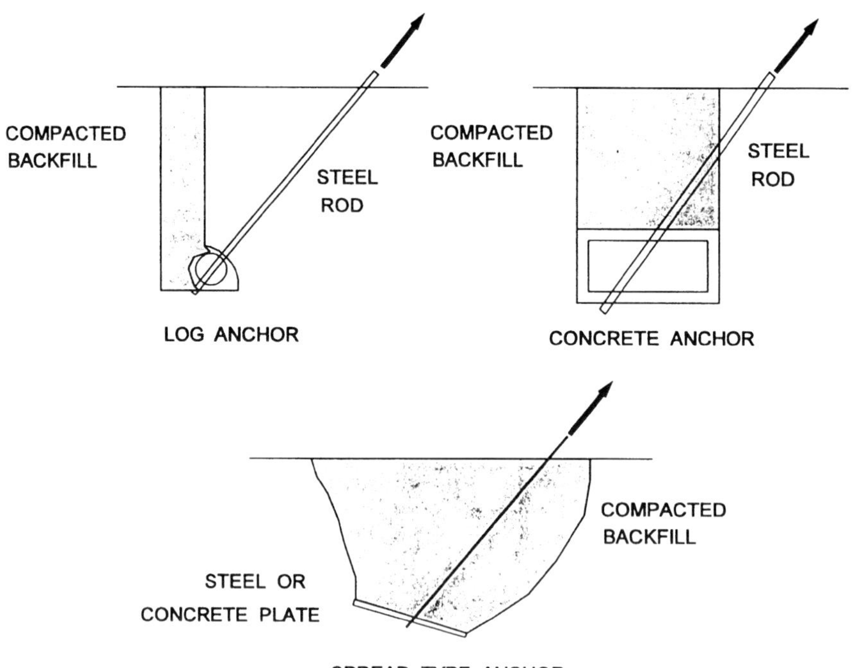

FIG. 4-1. Typical Deadman Anchor.

TABLE 4-1 Typical Holding Strength of Commonly Used Screw Anchors

Anchor		Holding Strength kN (kips)[b]	
Type	Size cm (in.)	Poor Soil	Average Soil
Screw or helix[a]	20 (8)	27 (6)	67 (15)
Screw or helix	29 (11-5/16)	42 (9.5)	67 (15)
Screw or helix	80 (32)	44 (10)	102 (23)
Screw or helix	86 (34)	69 (15.5)	120 (27)

[a]Screw or helix anchors power-installed.
[b]Values provided for illustration purposes only. Check with manufacturer for actual design values.

soil. The screw anchor develops its uplift capacity from the bearing capacity of each helix. In most soils, the capacity of these anchors can be related to the torque required to install the anchor, depth of installation, quantity of helixes, and the soil type. The use of torque for design should be confirmed by sample uplift testing. Manufacturers normally furnish charts that relate these parameters with the ultimate capacity. Table 4-1 gives some typical capacities of screw-type anchors and Fig. 4-2 shows typical examples.

4.3 GROUTED ANCHORS

Grouted anchors may be installed in either rock or soil. Typical grouted anchors are shown in Fig. 4-3.

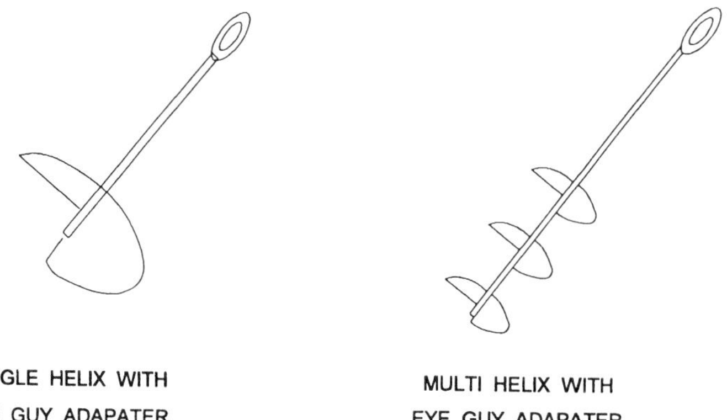

FIG. 4-2. Screw Anchors.

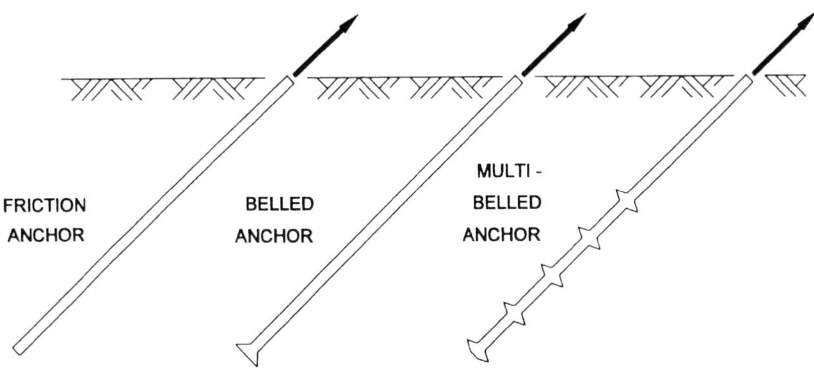

FIG. 4-3. Typical Grouted Anchors.

When installing anchor rods in rock, the diameter and length are determined by the strength of the rod, shear capacity between the grout and rod, shear capacity between the grout and rock, and capacity of the rock in shear. Normally anchor rods are standard reinforcing steel although high-strength steel bar stock is sometimes used. A hole is predrilled at the same angle as the guy. A steel rod is inserted in the hole and is then grouted. The hole size for rock anchors ranges from 1.5 to 3 times the diameter of the rod. Before grouting, the holes should be washed and water allowed to stand in the holes overnight. Prior to installation of the anchor bar, the holes are blown clean. Usually the grout is a water–cement mixture with an approved expansion agent added per the manufacturer's instruc-

TABLE 4-2 Typical Ultimate Bond Stresses Between
the Grout Plug and Rock

Rock Type	Ultimate Bond Stress[b]	
	kPa	psi
Granite and basalt	1700-3100	250-450
Dolomite limestone	1400-2100	200-300
Soft limestone[a]	1000-1500	150-220
Slates and hard shales	800-1400	120-200
Soft shales[a]	210-800	30-120
Sandstone	800-1700	120-250
Concrete	1400-2800	200-400

[a]Bond strength must be confirmed by pullout tests which include time creep tests.
[b]Values shown for illustration purposes only. Check geotechnical reference for actual design.

tions. The most common grout is a pure cement grout mixed in the ratio of 17 liters (4.5 gallons) to 400 N (90 lbs) of cement. The grout should be used within the first hour after batching. Typical ultimate bond stress values between the grout and rock for rock anchors are shown in Table 4-2.

Table 4-2 provides some guidance on ultimate bond stress, but it is strongly recommended that core drilling be performed to explore the rock quality, and core testing to determine the rock strength. In addition, after completing installation of the anchor rod conducting an anchor pull test to confirm the anchor capacity is recommended.

When grouted anchor rods are installed in soil, resistance is provided by the friction between the grout and soil as well as the bearing where anchors have bells with larger diameter than the initial shaft diameter as shown in Fig. 4-3. Grouted anchors in soil are very similar to those installed in rock. A water–cement grout mixture is inserted by gravity or under pressure. Injecting the grout will increase the capacity of the anchor and is dependent on the soil–grout interface. As with grouted rock anchors, it is recommended that proof testing be conducted after installation.

Chapter 5

ANALYSIS

This section presents theoretical and practical concepts for the analysis of guyed transmission structures. The concepts are further illustrated in the examples of Section 8. The term "analysis" in this section covers not only methods used to find forces, moments, displacements, and stresses but also methods for predicting the ultimate buckling capacity of the structure that often governs the design. Section 5.1 applies to all structures. Sections 5.2 and 5.3 deal more specifically with redundancy problems in poles and H-frames. Section 5.4 applies mostly to rigid frames and masted towers. Approximate manual procedures for predicting the ultimate buckling behavior of simple poles and masts are shown in Section 5.5. Finally, Section 5.6 discusses computer modeling.

5.1 CABLE BEHAVIOR

Unlike cables used as guys in tall communication structures (TV and communication antennae), guys in electric power structures are relatively short and usually taut. Their behavior is most often similar to that of straight bars of similar mechanical properties. It is important that the designer of guyed structures understand conditions under which the straight bar model is applicable.

The following geometrical and mechanical properties are used to characterize a guy cable.

$$h = \text{structure height to lower guy (m or ft)}$$
$$L = \text{chord length—from anchor to structure attachment point (m or ft)}$$
$$ALPHA = \text{guy slope—angle measured from horizontal to chord (degree)}$$

Δ = horizontal displacement of structure attachment point (mm or in.)

D = outside diameter (mm or in.)

A = cross-section area (mm^2 or in.^{2})

W = bare weight per unit length (N/m or lbs/ft)

RBS = rated breaking strength or ultimate tension capacity (kN or kips)

F_G = tension force in guy—assumed constant along guy (kN or kips)

HF_G = horizontal component of tension force in guy (kN or kips)

VF_G = vertical component of tension force in guy (kN or kips)

PF_G = pretension force—value of tension in unloaded structure expressed as a fraction of RBS (%)

E = guy effective modulus of elasticity (MPa or ksi)

K_G = horizontal guy stiffness—horizontal guy reaction HF_G caused by a unit horizontal displacement Δ (kN/m or kips/in.)

In transmission structures, the effect of temperature on guy tension is generally neglected as discussed in Section 5.2.2.2. Any wind and ice load on the guy itself is also generally ignored. The relationship between tension and elongation is assumed linear; that is, the guy effective modulus of elasticity E is assumed constant. If the guy cable were a straight steel bar, its modulus of elasticity would be that of steel, E_S = 200,000 MPa (29,000 ksi). However, because a guy is normally made of wound strands, its effective modulus of elasticity E is smaller than E_S. A value of 160,000 MPa (23,000 ksi) is often used (ASTM A475–89; ASTM A586–92; CAN/CSA-G12–92).

Consider the arrangement shown in Fig. 5-1 in which three guys come to a common attachment point D. Guy AD has a slope of 30 degrees (shallow), guy BD has a slope of 45 degrees, and guy CD has a slope of 60 degrees (steep). As the attachment point D is moved horizontally by the amount Δ, the guys can assume taut positions (Δ positive) or slack positions (Δ negative). The arrangement in Fig. 5-1 was used to generate the relationships in Figs. 5-2 and 5-3. These relationships are the basis for some concepts used in the analysis of guyed structures. The figures show relationships between the horizontal component of guy tension HF_G and the horizontal displacement Δ. They were obtained by using an exact cable element computer model (Peyrot and Goulois, 1979).

Figure 5-2 illustrates the behavior of short guy cables representative of what might be used to support a wood pole (height to attachment point h = 10 m (33 ft), $A \times E$ = 10,000 kN (2,250 kips), W = 4 N/m (.274 lbs/ft), RBS = 70 kN (15.7 kips)). Figure 5-3 is similar to Fig. 5-2 for longer guy cables representative of what may be used in a latticed tower where h = 30 m (99 ft). The properties for the guys in Fig. 5-2 were selected to be round numbers close to properties for

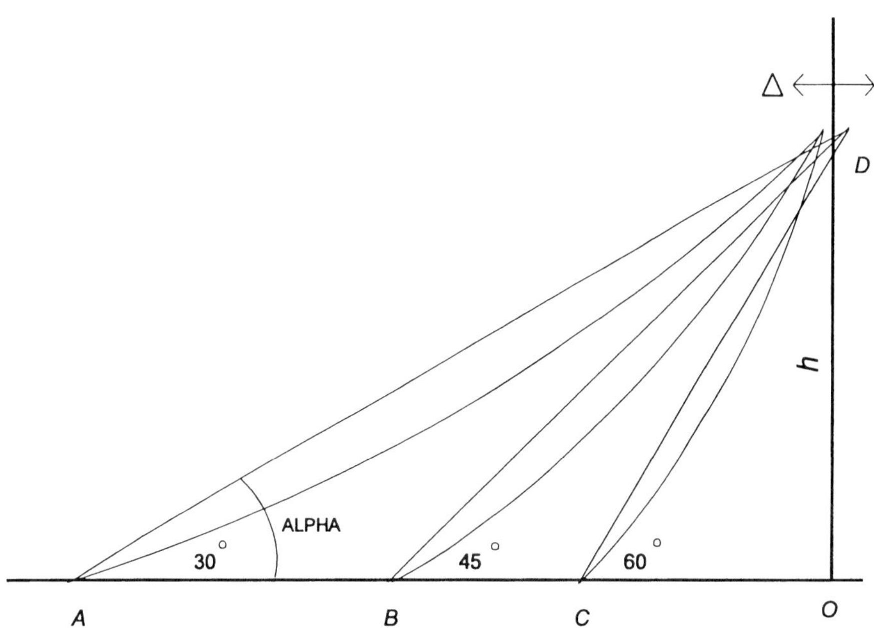

FIG. 5-1. *Three-Guy System in Slack and Taut Configurations.*

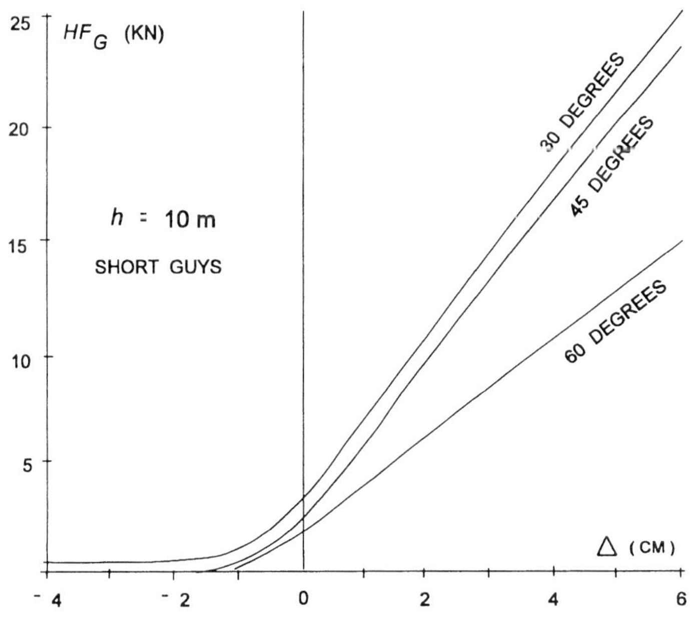

FIG. 5-2. *Tension Versus Displacement in Fig. 5-1.*

a 3/8 in. EHS cable ($D = 9.52$ mm). The properties for the guys in Fig. 5-3 are five times larger than those of the guys in Fig. 5-2. Although the curves were developed by assigning pretensions of 5% of *RBS* for all cables at no horizontal displacement ($\Delta = 0$), they can be used to study the effect of any pretension by treating Δ as a relative value.

For the short guys the leftmost portions of all curves indicate slack behavior (arbitrarily defined as a state of tension less than 1% of ultimate). The rightmost portions of all curves tend toward a straight line indicating taut behavior. Consider for example the 45 degree cable. With a pretension of 5%, a 2 cm (.79 in.) displacement to the right increases its horizontal component of tension from 2.47 to 9.52 kN (5 to 19% of *RBS*). A 2 cm displacement to the left decreases its horizontal tension from 2.47 to .22 kN, a completely slack condition. However, if the cable is preloaded at a horizontal tension of 9.52 kN, a 2 cm displacement to the left reduces the horizontal tension from 9.52 to 2.47 kN (19 to 5% of *RBS*): the pretensioned cable is still taut after a displacement to the left.

For the longer guys in Fig. 5-3 preloaded at 5% of their strength, the transition from taut to slack is not as rapid as for small guys. A 2 cm displacement to the left reduces the horizontal tension of the 45 degree guy from 12.4 to 6.1 kN (5 to 3% of ultimate).

The slope of any curve in Fig. 5-2 or 5-3 represents the instantaneous value of the horizontal guy stiffness K_G. In all cases (small or

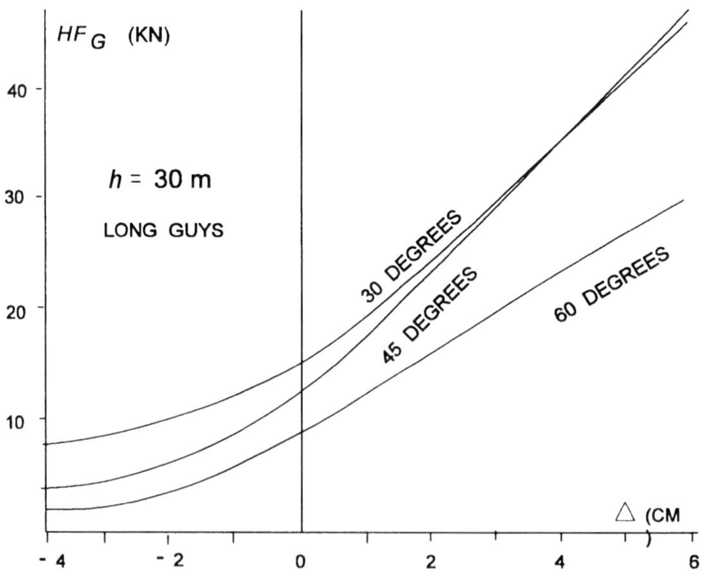

FIG. 5-3. *Tensions Versus Displacement for Guys in Fig. 5-1.*

large guys), and for tensions larger than 5% of *RBS*, the behavior is almost linear with stiffness close to that of an equivalent straight inclined bar:

$$K_G = A \, E \cos^2[ALPHA]/L. \qquad (5\text{-}1)$$

The preceding discussion provides justifications for the simplified modeling of guys by straight tension-only bars. The approximation is better for smaller structures. For structures with guy lengths in excess of 60 m (about 200 ft), the designer should consider using a suitable nonlinear cable element with a nonlinear computer analysis.

5.2 POLES OR LATTICED MASTS WITH SINGLE GUY ATTACHMENT POINT

This section discusses simple concepts applicable to single guyed poles or masts, regardless of material. The poles or masts are vertical and capable of resisting both axial compression and bending moment.

5.2.1 Single Guy Level and Hinged Base

Figure 5-4 shows a guyed pole with a single back guy subjected to a transverse force *H*. Because the pole is hinged at the base, the entire transverse force is resisted by the horizontal component of guy tension; that is, $HF_G = H$. The vertical component VF_G follows from the guy slope. That component creates compression in the pole, $VF_P = VF_G$.

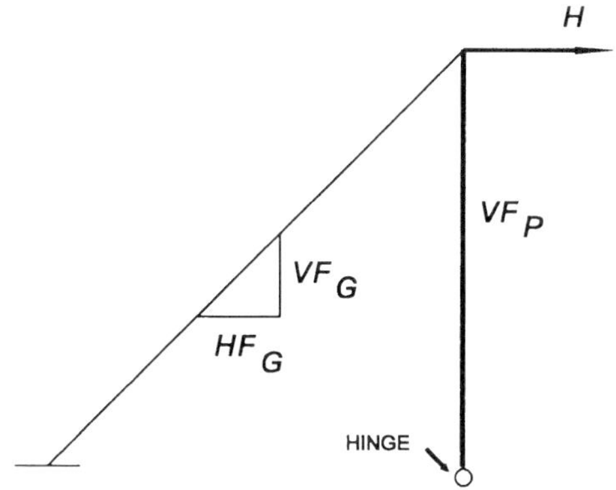

FIG. 5-4. Guyed Mast.

In Fig. 5-5 the transverse force is resisted by a pair of back guys. The top view of the system shows how each guy shares the transverse load equally. The force in each guy can be obtained by similar prisms: $F_G = (H/2) L/t$, where t, l, and h are the transverse, longitudinal, and vertical projections of the guy, and L is the guy length.

In Fig. 5-6 the pole is held by a pair of 45 degree prestressed guys, a back guy AC, and a head guy BC. Assume that the guys are prestressed at 14.14 kN. This pretension causes an axial force of 20 kN in the pole. When the transverse load is increased from zero to 30 kN, the tension in the back guy increases from 14.14 to 42.43 kN, whereas that in the head guy decreases from 14.1 kN to zero. The head guy becomes slack for loads larger than or equal to 20 kN. For a load of 30 kN, the back guy tension is 42.43 kN, exactly the same as if there had been no prestress. From this it may be noted that as long as a guy prestress does not exceed 50% of the maximum loaded guy tension, the prestress will not have an effect on the maximum tension. Before the head guy becomes slack, the guy tensions are simply calculated as the superposition of the initial guy preloads and the forces in the truss made up of the mast and two straight bars replacing the guys.

When a hinge exists or is assumed to exist at the base of a pole, the corresponding model is referred to herein as a ''column or strut'' model and the analysis as a ''column or strut'' analysis.

5.2.2 Single Guy Level and Fixed Base

Figure 5-7 shows a pole fixed at its base and supported by a single back guy. The system is now statically indeterminate. The transverse load H is shared between the horizontal component of guy tension

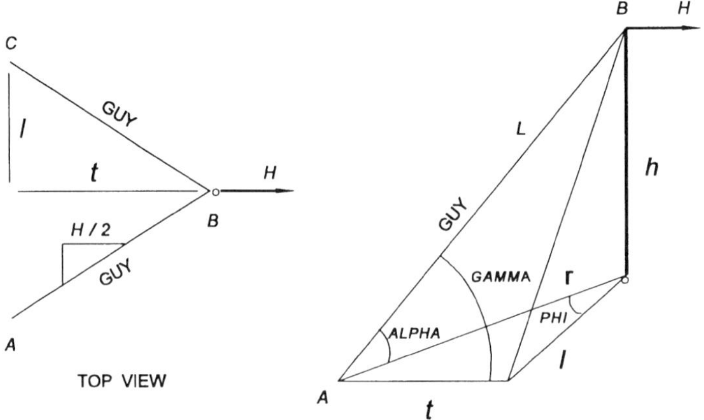

FIG. 5-5. *Hinged Mast with Two Guys.*

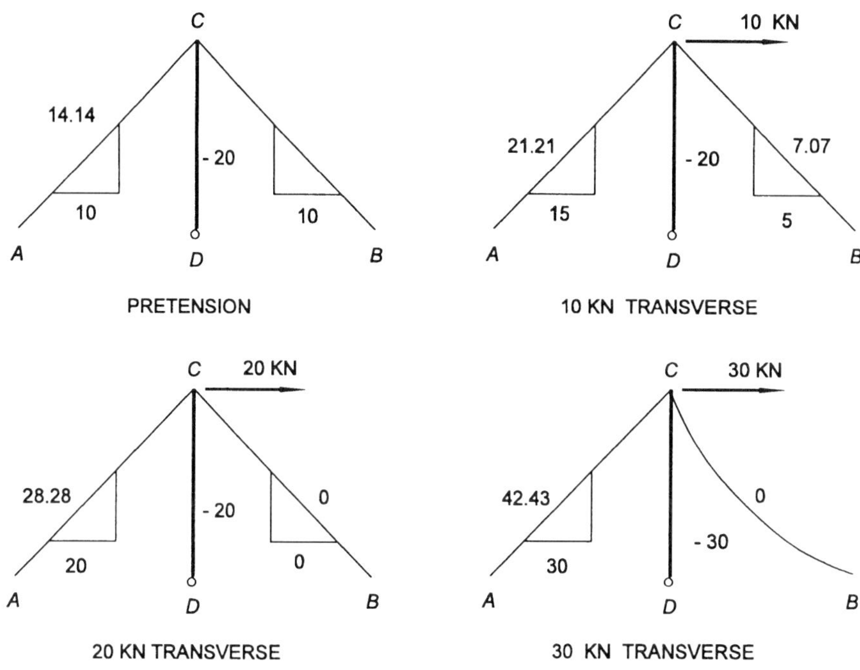

FIG. 5-6. *Hinged Mast with In-Line Back and Head Guys.*

HF_G and the shear in the pole HF_P; that is, $H = HF_G + HF_P$. The relative magnitudes of the forces HF_G and HF_P depend on the horizontal stiffnesses K_G and K_P of the guy and the pole, respectively. The situation is similar to that of two springs sharing the load. It can easily be shown that:

$$HF_G = H \times K_G/(K_G + K_P) \quad \text{and} \quad HF_P = H \times K_P/(K_G + K_P). \qquad (5\text{-}2)$$

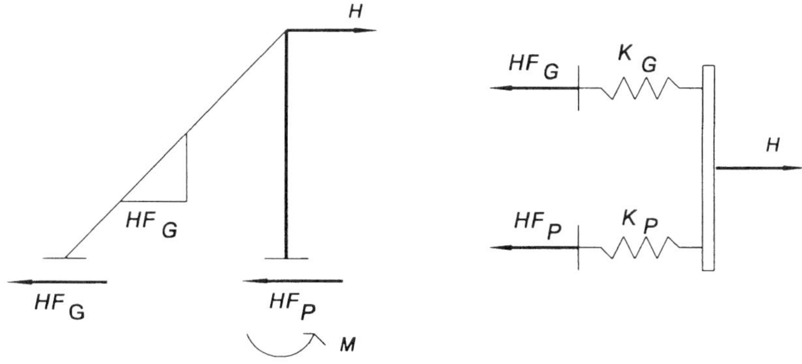

FIG. 5-7. *Fixed Base Pole with In-Line Guy.*

The ratios $K_G/(K_G + K_P)$ and $K_P/(K_G + K_P)$ are distribution factors distributing the transverse load to the two components of the system. The moment at the base of the pole is simply $M = h \times HF_P$.

For a constant cross-section pole of height h, modulus of elasticity E_P, and moment of inertia I_P:

$$K_P = 3 \times E_P \times I_P/h^3. \tag{5-3}$$

For a tapered wood pole of height h, modulus of elasticity E_P, top diameter d_a, and groundline diameter d_g:

$$K_P = E_P \times d_g^3 \times d_a/(6.79 \times h^3). \tag{5-4}$$

Consider two 12 m (39 ft) poles, one made out of steel and the other of wood. Both are guyed with a single back cable having the realistic properties:

$$A \times E = 10{,}000 \text{ kN } (2{,}250 \text{ kips}),$$
$$L = 12 \times 1.414 = 17.0 \text{ m } (55 \text{ ft}),$$
$$ALPHA = 45 \text{ degrees}.$$

From Eq. (5-1), $K_G = 294$ kN/m.

The steel pole has a 50 cm (19.7 in.) outside diameter, a thickness of .5 cm (.197 in.), and $E_P = 200{,}000$ MPa (29,000 ksi). Its moment of inertia can be calculated to be $I_P = .000238$ m^4. From Eq. (5-3), $K_P = 82.6$ kN/m. If there were any base rotation, K_P would be smaller.

The wood pole has top and groundline diameters of 25 (9.8 in.) and 50 cm (19.7 in.), respectively. Its modulus of elasticity is $E_P = 14{,}000$ MPa (2,030 ksi). From Eq. (5-4), $K_P = 37.3$ kN/m. In reality, because of base rotation and creep deflection, the long-term value of K_P would be much smaller than 37.3 kN/m.

From the preceding, it can be concluded that the guy in the steel pole example carries $294/(82.6 + 294) = 78\%$ of the transverse load. The remaining 22% becomes shear in the pole, causing base moment. In the wood pole example, the guy theoretically carries $294/(37.3 + 294) = 89\%$ of the load. Actually, after base rotation and creep, the guy carries almost all the load.

If the pole in Fig. 5-5 were fixed at the base, the transverse load would be distributed among the pole and the two guys. Instead of two springs in parallel, we would have three springs in parallel. The horizontal stiffness of each guy would be obtained with Eq. (5-1) with the angle $ALPHA$ replaced by the angle $GAMMA$ shown in Fig. 5-5.

Referring to the guy arrangements in Fig. 2-2 for a single strain pole located at a small line angle (line angle $= 2\,PHI$), the horizontal stiffness of an in-line guy in the transverse direction is (from Eq. (5-1)):

$$K_G = A\,E\,\cos^2[\cos^{-1}\{\sin(PHI)\cos(ALPHA)\}]/L. \qquad (5\text{-}5)$$

From Eq. (5-5), assuming a guy slope $ALPHA$ = 45 degrees and a line angle of 10 degrees (PHI = 5 degrees), the stiffness of an in-line guy is .0038 $A\,E/L$ and that of a bisector guy is .5 $A\,E/L$; that is, the bisector guy is 131 times more efficient in carrying load than the in-line guy. This is the reason why, for small line angles and intact loading (pull in the direction of the bisector), in-line guys are not effective. For small line angles and intact loads, modified in-line guys (moved toward the bisector as shown in Fig. 2-2(b)) or bisector guys (Fig. 2-2(c)) are more effective. For poles that have to support both intact and dead-end loadings, the combination of in-line and bisector guys, as shown in Fig. 2-2(d), is the best solution: the in-line guys take care of the dead-end loads and the bisector takes care of the intact load.

In summary, the amount of base moment in a pole guyed at a single level depends on the relative stiffnesses of the guys and the pole. Because directly embedded wood poles are generally very flexible, it has traditionally been assumed that all the transverse load is picked up by the guys, leaving no moment at the base of the pole. With this ''column'' analysis, the wood pole acts as a pure compression member which is only checked against buckling. Directly embedded or fixed base steel or concrete poles may be stiff enough to attract some lateral load and develop significant base moments. The amount of transverse load carried by the pole depends on the relative stiffnesses of the guys and the pole. For tapered steel or concrete poles, the determination of the stiffness of the pole K_P is not practical without the use of a computer program.

The problem of determining the proper sharing of load between the guys and a fixed base pole is the same as the load-sharing problem in a guyed rigid base latticed tower. The danger in such systems is that the distribution of load may change with time as the foundations move or the various materials creep. This is briefly discussed in the following.

5.2.2.1 Effect of Preload or Foundation Movement. All the previous calculations for fixed base poles assume that there is no preload at zero deflection; that is, the pole is perfectly plumb and the guys are slack at installation. However, a guy may be installed with a pretension, the horizontal component of which is HF_{G0} as shown in Fig. 5-8. The pretension may also be affected by movement of the anchor. The pretension creates a base moment M_0 at the base of the pole, where $M_0 = h \times HF_{G0}$ (ignoring the P-Delta effect). If the pretensioned system is then loaded with a transverse force H, the forces and moments in the final equilibrium position are the algebraic superposition of forces and moments in Figs. 5-7 and 5-8. The final guy force has a horizontal component which is the sum of HF_G in Fig. 5-7 and HF_{G0} in Fig. 5-8. The final base moment is the difference between M and M_0. This example points to the fact

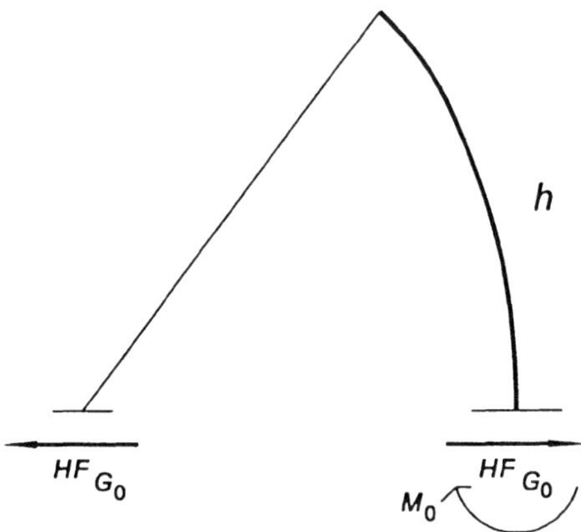

FIG. 5-8. Guy Pretension.

that, for fixed base poles or rigid towers, the guy installation procedure (preload) or foundation movement should be known or assumed before an analysis is made.

Foundation movement, whether translation or rotation, can also significantly affect the analysis results.

5.2.2.2 *Effect of Temperature.* For the simple examples of this section as well as actual guyed transmission structures, temperature has a negligible effect on guy tensions. This is particularly true when both guys and structures are made of steel.

5.3 POLES OR LATTICED MASTS WITH MULTIPLE GUY ATTACHMENT POINTS

Transmission poles are usually guyed at several levels. Using guys at each transverse load point will minimize the amount of bending in the nearby portion of the pole.

5.3.1 Multi-Guy Levels and Hinged Base— "Column" or "Strut" Model

If there is a guy at each transverse load application point and if the pole is hinged at each guy level and at ground line (Fig. 5-9), then all transverse loads are carried by the guys. The pole resists the sum of all vertical components of guy tensions. When hinges exist or are assumed to exist at the pole base and all guy levels, a "column" or "strut" model is used.

The "column" assumption is generally not accepted for steel and

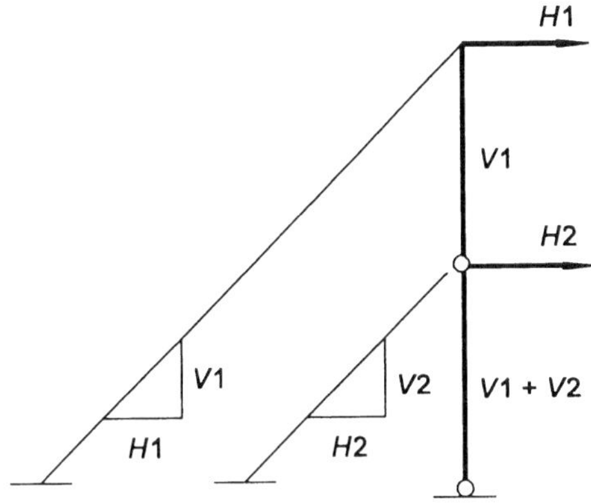

FIG. 5-9. Hinged Mast.

concrete poles, even when they are hinged at their base, because of their greater stiffness and lower load or safety factors.

Although wood poles guyed at several levels are actually not hinged at those levels and are certainly not hinged at their base, they have traditionally (IEEE 1991; REA 1982) been assumed sufficiently flexible to snake through the guy attachment points without developing serious moments. Therefore traditional guyed wood pole analysis uses the "column" assumption to compute the loads in the guys and the axial force in the pole. At the design stage, the wood pole is sized so that its buckling compression capacity exceeds the axial force.

5.3.2 Multi-Guy Levels and Fixed Base

Because fixed base poles with multiple guy levels are highly indeterminate structures with potential for buckling, they should be analyzed by geometrically nonlinear computer programs. Use of the "column" model (traditionally accepted for wood poles) can be misleading, as demonstrated by the example in Section 8.2.1.

5.4 STRUCTURES WITH FOUR GUYS

Approximate manual methods for calculating guy tensions in structures with four guys can be used. These methods require that some assumptions be made regarding which guy or guys are slack under a specific load case. The slack guys are simply removed from consideration.

Figure 5-10 shows top views of typical four-guy patterns. Thick solid lines indicate which guys are expected to be taut under the two

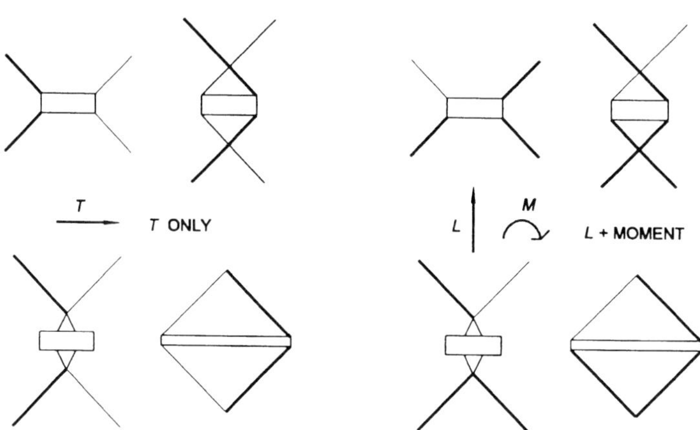

FIG. 5-10. *Taut (Thick) and Slack (Thin) Guys.*

load cases shown: pure transverse load T and broken outer phase conductor load equivalent to a combination of longitudinal force L and torsional moment M at the center of the structure. Thin lines identify the slack guys that are ignored.

With two or three active guys, it is often possible to determine all guy forces and structure reactions by using basic principles of statics. There are many ways of approaching a problem, depending on the geometry of the structure itself, however, there are typical steps that are often used and which are presented in the example section (Section 8).

5.5 BUCKLING STRENGTH OF POLES AND LATTICED MASTS

5.5.1 Pole Buckling Strength

Guyed poles are subjected to large compression loads which may lead to buckling failure. Therefore a failure analysis which is capable of predicting large displacements and buckling is normally required.

Well-known analytical formulae exist for simple cylindrical poles with length h, modulus of elasticity E_P, and moment of inertia I_P. For the four poles in Fig. 5-11, if $d_g = d_a$, the theoretical buckling load P_{CR} is:

$$P_{CR} = \pi^2 E_P I_P/(K h)^2, \tag{5-6}$$

where the equivalent pin-end length factors K are as shown. The K factor in Fig. 5-11(d) is for a pole with a single in-line back guy where the compression force in the pole is not vertical as in Fig. 5-11(b), but always passes through the base.

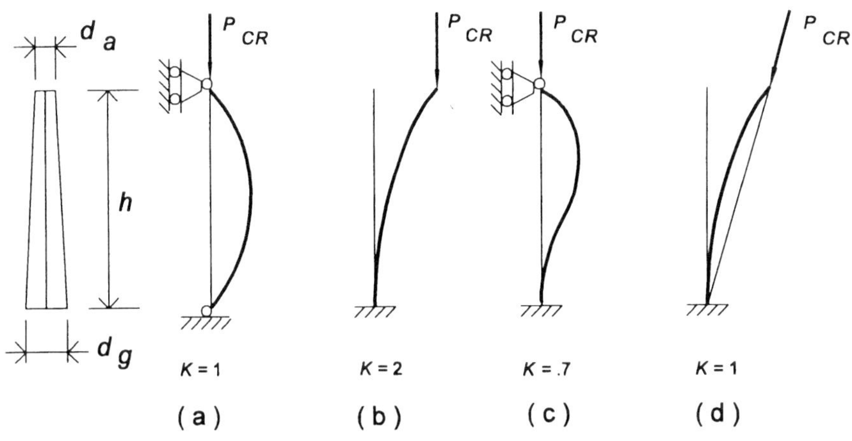

FIG. 5-11. *Buckling Shapes ($d_a = d_g$).*

For simple tapered poles, Eq. (5-6) can be replaced by (Gere and Carter 1962):

$$P_{CR} = P^* \pi^2 E_P I_a / (K h)^2, \qquad (5\text{-}7)$$

where I_a is the moment of inertia at the top of the pole (based on d_a) and P^* is a correction factor that is a function of the diameter ratio d_g/d_a and the moment of inertia ratio I_g/I_a. Curves and formulae are included in the Gere and Carter paper for a variety of cross-sectional shapes and end restraint conditions. For wood poles with a solid circular cross-section, the correction factor is equal to $(d_g/d_a)^2$ for the conditions in Fig. 5-11(a), (c), and (d). It is approximately equal to $(d_g/d_a)^{2.7}$ for the condition in Fig. 5-11(b).

It should be realized that the simple poles for which analytical formulae exist (as in Fig. 5-11) must have at each end one of four ideal conditions: perfectly fixed, pinned with lateral displacement prevented, free to move in a direction perpendicular to an in-line back guy, or free. In real poles, these ideal conditions rarely exist. A strain pole with four in-line guys at a 90 degree line angle will likely buckle with the shape shown in Fig. 5-12. The condition at the lowest conductor is neither pinned nor laterally fixed as is usually assumed for wood poles.

Although the formulae of Eqs. (5-6) and (5-7) are often used for wood poles (IEEE 1991), they can lead to large errors as shown for the wood pole examples of Section 8. Simple buckling checks with Eqs. (5-6) and (5-7) do not cover cases where the ultimate capacity of the pole is controlled by bending stresses amplified by large compression load.

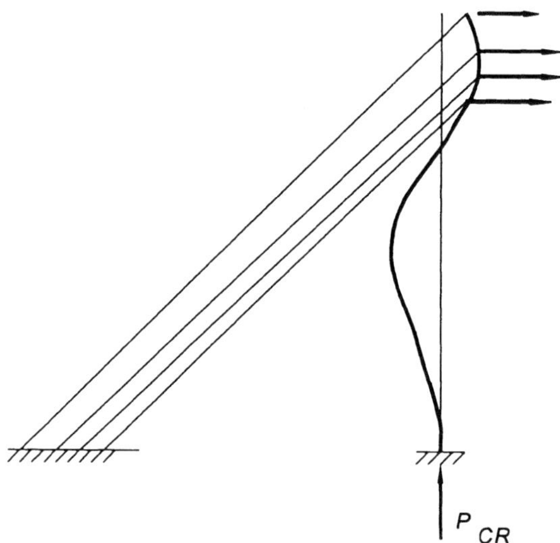

FIG. 5-12. Buckled Shape.

For tapered steel and concrete poles with nonuniform wall thickness, there are no simple buckling formulae. However, the buckling analysis can be done numerically with the appropriate computer program. Concrete poles are seldom controlled by buckling.

5.5.2 Equivalent Beam Model for Latticed Masts

Latticed masts in a guyed structure can be replaced by series of equivalent beams. This is required for a manual analysis and it can also be done for a simplified computer analysis. The alternate is to model all the members in the mast, but that requires a more complex computer analysis.

Figure 5-13 shows a typical latticed mast with square or rectangular cross-section. Triangular cross-sections can also be used. The main loading on the mast is its axial load P (kN) and lateral or quartering wind load w (kN/m). The axial load P is obtained from an overall analysis of the entire guyed structure. In addition to P and w, there might be small transverse and longitudinal end moments, as well as mid-mast moments caused by out-of-straightness. Out-of-straightness due to bolt slippage during lifting of the mast is not a problem. Under full load (or test), the bolts will slip in any event and thus deflection from bolt slippage, whether from erection or load, must be assumed and added to elastic deflection due to wind on the mast to become the important lateral deflection (delta) value. An out-of-straightness of between $L/200$ and $L/500$ is often assumed.

The mast can be analyzed as a beam-column simply supported at

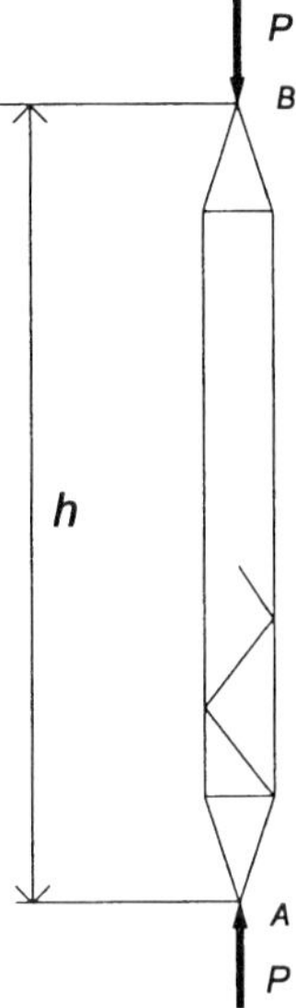

FIG. 5-13. Mast.

its ends A and B. The moments obtained using a first order linear analysis that ignores the effects of deflection (also called primary moments) should be amplified by the factor:

$$1/(1 - P/P_{CR}), \tag{5-8}$$

where P_{CR} is the buckling load of the mast. Ignoring the effect of end tapers, the buckling load can be calculated as:

$$P_{CR} = CF \times (\pi^2 E_M I_M / h^2), \tag{5-9}$$

where:

$E_M I_M$ = modulus of elasticity times mast moment of inertia
CF = correction factor to account for the fact that shear in the bracing increases deflections beyond what is caused by bending only. Detailed expressions for CF in terms of the slope of the bracing, mast width, and areas of diagonal and strut members can be found in steel design textbooks (Bresler et al. 1960; Johnson 1976; Salmon and Johnson 1990). A reduction of 5%, (i.e., CF = .95) has been used.

Moments and shears are finally resolved into individual member axial forces.

5.6 COMPUTER MODELING

Guyed structures, as compared to self-supporting structures, present unique analysis problems. In addition to the geometrically nonlinear behavior of the guys (modeled as tension-only members or true cable elements), one should take into account the effects of potentially large deflections and large axial compression loads in slender members or masts which can create significant P-Delta effects and lead to buckling. For reasons given in the following, it is strongly recommended that a geometrically nonlinear computer analysis be used to determine final design forces, moments, and capacity in guyed structures, including wood poles and wood H-frames for which safety factors do not already account for possible nonlinear effects.

5.6.1 Linear Computer Analysis

If a linear computer analysis or a manual analysis is used, the guys can be modeled with tension-only straight bars. This requires that each load case be handled separately after removing the bars that are under compression. Preload can be accounted for. A design process that relies on a linear computer analysis or a manual analysis requires further work beyond that analysis. The linear analysis produces primary (or first order) values of forces, moments, and deflections. Since these primary values do not include their possible amplification due to overall structure displacements or due to local flexing of slender compression subsystems (pole or mast) between braced points, they may have to be amplified. Finally, one should determine that the entire structure or any one of its parts will not buckle under the amplified forces and moments. As discussed in Section 5.5, it may be difficult to perform an accurate buckling check as analytical expressions for buckling loads are seldom available.

For example, when imposing a broken wire case on a guyed Delta (Section 2.5.3), the structure rotation from a linear or manual analysis does not include the geometrically nonlinear effects due to the displacements of the vertical and transverse loads. Actual guy loads and, subsequently, mast loads can be 10 to 20% higher due to the P-Delta effects. It also is important to note that the pretension value used in the analysis will also affect the amount of mast rotation under these conditions.

For laced masts, a simple method to account for within-mast moment amplification was presented in Section 5.5.2. Amplified quantities are then used to verify that corresponding stresses and displacements are below allowable values.

Approximate manual methods or linear computer methods may be used for certain types of structures (simple wood poles and masted towers) as long as these methods are accepted by governing codes, for ''sanity'' checks of computer solutions and better understanding of how a guyed structure works, and at a preliminary design stage.

5.6.2 Nonlinear Computer Analysis

A geometrically nonlinear or second order analysis (the recommended method) should compute member forces and moments that are in equilibrium in the displaced structure configuration. It should be capable of predicting elastic instability phenomena by showing, as in Fig. 5-14, increasingly larger deformations and stresses as the loads approach buckling. Therefore, with a geometrically nonlinear analysis, there is no additional work required beyond the analysis as is the case with a linear analysis. If an analysis has converged to an equilibrium configuration where displacements and stresses are acceptable, then the design is structurally acceptable.

With the appropriate software (nodes and members of an analysis model are generated automatically from a few key structural dimensions and material properties), a nonlinear analysis is not only a trivial task on current microcomputers but it is also a cost-effective and reliable way to produce a design. Even for guyed wood poles and H-frames, the nonlinear analysis relieves the designer from having to select among many unproven and often conflicting buckling formulae.

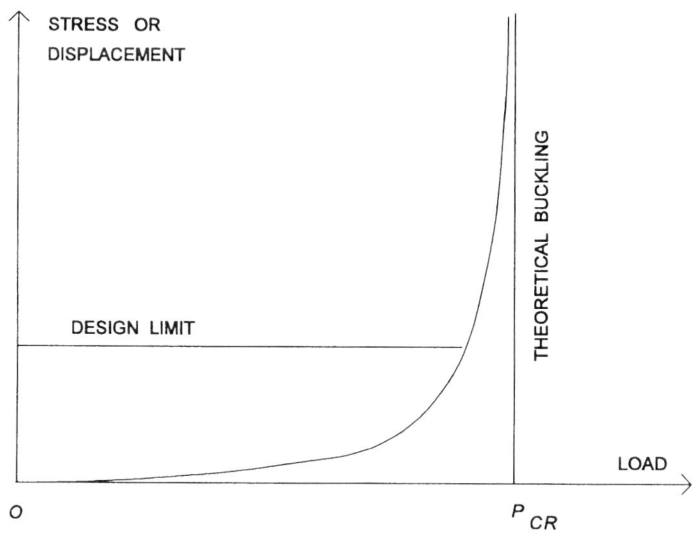

FIG. 5-14. *Stress-Load Relationship up to Buckling.*

5.6.3 Modeling Hints and Details

The attachment point of a guy to a pole is eccentric with regard to the pole center line. A computer model often assumes that the guy is attached to the point at the same elevation on the center line. If this is the case, it should be realized that the effect of the eccentric moment caused by the guy tension is not taken into account. If eccentricity is a concern, there are several ways to account for it: (1) if the attachment point elevation is fixed, the center line point can be raised to the intersection of the guy line with the center line (Fig. 5-15(a)); (2) if the elevation at the center line is prescribed, the attachment point of the guy can be lowered at the detailing stage (Fig. 5-15(b)); or (3) a rigid link can be used in the analysis to connect the center line point to the attachment point (Fig. 5-15(c)).

The support condition at the base of a real pole ranges from pinned (if rotation occurs freely) to fixed (if the base is rigid). Either of these extreme conditions is easy to model on the computer. For an intermediate situation some limited rotation occurs. To account for this effect some analysts assume that the pole is not fixed at ground line but at a certain distance below (e.g., .5 m or 1/3 setting depth). Others insert a rotational spring at the base, with properties coming from the moment-rotation relationship of the foundation. A third alternative is to model the pole and its foundation as a unit. In general, because of the uncertainty regarding the behavior of the foundation, there is little justification for using a complex spring or foundation model. An effect of groundline fixity is illustrated in Section 8.2.2.

When stub poles are used, it is recommended that the complete system consisting of the primary pole and all its guys and stub poles be analyzed as a single unit with a nonlinear computer program.

Since transmission structures are not normally designed for earthquakes and since equivalent static loads are used to model other dynamic effects (broken conductors, span galloping, etc.), dynamic analysis is not usually considered for guyed transmission structures.

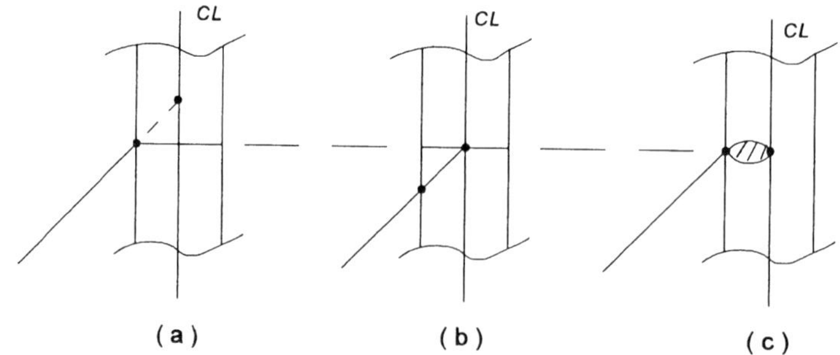

FIG. 5-15. Eccentric Guy Connection.

5.6.3.1 *Special Considerations for Guyed Concrete Poles.* As was discussed in Section 5.2.2, the distribution of lateral load between a pole and its guys depends on their relative stiffnesses. For concrete poles, the stiffness depends on an assumed $E\,I$ (modulus of elasticity times moment of inertia). The forthcoming *ASCE Guide for the Design of Prestressed Concrete Poles* recommends $E\,I$ values ranging from $E_c I_c$ for uncracked poles to $(E_c I_c)/3$ for cracked poles near an ultimate bending condition, where E_c is the concrete modulus of elasticity and I_c is the gross uncracked cross-section moment of inertia. Since guyed concrete poles generally remain uncracked over most of their length, their analysis is more appropriately made with $E\,I = E_c I_c$. For unguyed poles near failure in bending, the $E\,I = (E_c I_c)/3$ value is more appropriate.

Chapter 6

DESIGN

6.1 LOADINGS AND DESIGN RESTRICTIONS

Like any other transmission structure, a guyed structure has to meet as a minimum the requirements of the governing code (NESC 1993 or current). However, other weather-related criteria and special loads should also be considered. These additional loads may include loads from special construction techniques, loads from guy pretensions, and longitudinal loads to prevent cascading. General guidance for the selection of loads is provided by the ASCE *Guidelines for electrical transmission line structural loading* (ASCE 1991).

In guyed structures, the length of the guys is usually such that wind and ice loads on the guys may be neglected. Although ice buildup on members of self-supporting latticed transmission towers is normally neglected, ice buildup on tightly latticed masts might be considered in design. This is due to the fact that a tightly latticed mast can have its solidity ratio significantly altered by the ice condition and the wind-on-ice together with the ice weight can result in critical bending moments in the mast. The designer should also be aware that oblique winds on a mast may be the critical loading condition.

In very cold climates frost heave may jack up the mast footing but may not affect the guy anchors that are set below the frost level. This may increase the guy loads and the total load may crush the tower. A structural fuse in one of each cluster of four guys will remove the possible problem. A similar problem may be caused by expansive clay.

Some utilities require that fixed base poles and H-frames be designed to withstand an everyday loading or higher without any guy wires attached.

Because of their importance and relatively low costs, guys should not be the weak link among all the components of a structure. A coordination of strength (ASCE 1991) according to which the guys, their fittings, and anchors are more reliable than other structural components can be achieved with the appropriate use of strength or load factors.

The following basic guy restrictions should be determined by the transmission line designer in conjunction with the structure designer prior to beginning design of the structure.

1. common guy sizes typically used
2. maximum and minimum guy slopes
3. whether multiple guying is acceptable
4. whether multiple guys should be attached to the same anchor
5. minimum guy anchor spacing
6. maximum guy force limitations due to guy anchors and hardware
7. special electrical clearances
8. site characteristics (rugged, farming, etc.)

6.2 GUY CLEARANCES

Guyed structures may be flexible enough to require clearance checks within the structure in the deflected shape. In particular, sustained longitudinal loads may significantly distort the structure. Dependent upon service requirements, both electrical and mechanical clearances must be maintained.

6.2.1 Electrical Clearances

Minimum electrical clearances for safety considerations are given in governing codes or utility criteria. Minimum required clearances from conductors to guys (rod to rod), from conductors to narrow cross-section masts (rod to mast), and from conductors to parts of towers (rod to plane) or poles may be different. From laboratory tests showing that ''rod to rod'' and ''rod to mast'' air gap resistances were substantially higher than ''rod to plane'' values, conservative air gap reductions of 15% for conductors to guys and 7% for conductors to masts, compared with gaps to rigid towers, have been used. Flashovers to guys, very seldom if ever recorded except when stuctures are brought down by a tornado or the equivalent, have never been known to burn through or significantly damage a guy.

6.2.2 Mechanical Clearances

The possibility of contact between crossing guys, except under broken conductor load, should be avoided. For Portal, Delta, or other structures that use the cross guy configuration the guy anchors should be offset in such a way as to avoid rubbing contact between the guys.

6.3 GUY DESIGN

6.3.1 Guy Pretension

The pretensions of mutiple guys, attached to rigid or hinged single poles of steel or concrete, are of critical importance in determining the behavior of the combined structural system. The influence of guy pretensions in these situations can only be explored by computer analysis, as described in Section 5.

Wood poles have so much inherent flexibility that small differences in guy length (and thus tensions) have very little effect on the distribution of stresses between the guys and the pole. Thus guys for wood poles may be pulled up tight and fitted directly to the anchor with no threaded in-line adjustment.

For all but one type of guyed structure (the exception being the guyed V with haunches), the amount of pretension in the guys will have little effect on loaded capability unless the pretension is extreme and exceeds 50% of the design maximum tension.

The designer should specify, or be aware of, the guys installation procedure as described in Sections 7.3.2 and 7.4.4. Depending on the procedure, the designer should specify either: (1) pretension values, (2) a tensioning sequence controlled by structure top displacement, or (3) as a minimum, that the guys should be tensioned "snug tight" at a given stage of construction.

For a tangent structure, the loads in all four guys will be equal so that there is no need to monitor the load in more than one guy. For these structures, pretensions typically range from 5 to 10% of Rated Breaking Strength (RBS). For practical considerations, the minimum pretension should be such that the leeward guys do not go slack under frequently occurring winds (say, yearly wind) or other load combinations. A tolerance of ± ten percent (± 10%) of the specified pretension is considered practical during construction.

Under normal service conditions guys usually will not stretch or loosen and therefore any detectable slackening could be the result of anchor movement. The initial tensioning during tower erection should take out any anchor set so that significant guy slackening could be a sign of a problem anchor. In this regard, guyed structures offer a warning system of possible anchor deficiency not available with other rigid types. Utilities are usually quick to monitor guy tensions for a year or so but experience will reduce inspection of guys to the same frequency as that given to any other structure type or component.

With the CRS type of structure (Fig. 2-12), the guy system is tensioned at erection against the construction or spacer rope and only sufficiently to ensure the correct mast top spacing. Once the line is strung, the weight of the conductors dictates the everyday guy loads.

6.3.2 Allowable Tensions

Guy stress under service loads such as extreme wind, wind and ice, and extreme ice is usually limited to 65% of the guy RBS to prevent exceeding the elastic limit of the guy. This limit is to ensure that the tension does not reach the yield point of the stranding and produce permanent set that would require retensioning the guys. This limit also gives a cushion or margin to the terminal devices which frequently do not reach the RBS of the guys when field installed.

Some line designers will allow the guy (and fitting) tensions to reach 85% of RBS under longitudinal or failure containment loading that results from a failed tower or broken conductor. The reasoning is that with some kind of failure at an adjacent tower, or span, the tower in question and its guys are going to be checked for damage and guy tensions adjusted as necessary. An allowable 20% increase in guy tension from longitudinal loads will permit a better deployment of the guys and anchors for the critical transverse wind loads. This can be a substantial improvement in design efficiency. The *National Electric Safety Code* (NESC 1993 or current) allows guy tensions to reach 90% of RBS when the structure is loaded with its designated factored loads.

Guy fittings and hardware were described in Section 3. Usually they will not be able to develop the full RBS of the guy but will be adequate if the guys are not loaded to about 70% of RBS under weather loads and 85% of RBS under failure containment conditions. The mechanical efficiency of guy fittings may be defined as the percent of the guy RBS the guy fitting is capable of developing. The efficiencies of some typical connections (Refer to sketches in Fig. 3-2) are given in Table 6-1. The values (AISI 1981) are approximate and are provided for general information only.

6.3.3 Guy Slope

Guy slope is usually defined as the angle in degrees between horizontal and the guy or as the ratio V/H of elevation to horizontal projection. As shown in Section 5.1, the guy effectiveness decreases for the steeper slopes (because of loss of guy stiffness and increases in both guy tension and resulting structure compression) and also for the shallower slopes (because of loss of stiffness and increased right-of-way cost from the additional guy length). Therefore an optimum guy slope can be found by a trial-and-error process. Guy slopes

TABLE 6-1 Efficiency of Wire Rope Connections

Type of Connection	Efficiency (%)
Zinc poured socket properly attached	100
Compression fittings properly attached	100
Preformed grips	Check with supplier
Wedged sockets	75 to 90
Cable clips	80
Plate clips—three bolts type	80
Splice eye and thimble:	
1/4 in. and less	90
5/16 to 7/16	88
1/2	86
5/8	84
3/4	82
7/8	80

in the range of 45 to 60 degrees from horizontal are common. An example of the effect of guy slope is included in Section 8.

6.4 GUY ANCHORAGE

Typical guy anchorages have been described in Section 4. The geotechnical design of these items can be based upon provisions of appropriate geotechnical references (IEEE 1985). Since it is difficult to theoretically determine a guy anchor creep deformation and corresponding capacity, these items are often individually proof tested during installation to guarantee their performance under different types of loading events. This proof testing also sets the anchor, ensuring that creep of the anchor does not cause excessive loss of prestress in the guy.

For angle and dead-end structures, the designer should be aware of the duration of the potential guy loads. Guy loads from everyday tensions are permanent in nature and should be matched with long-term strength properties of the anchors. Loads from extreme ice may last a few hours and those from extreme wind, a few seconds. For these loads of shorter duration, the anchors will normally creep less and have greater strength.

One of the advantages of guyed structures is that guy anchor locations need not be determined with the same degree of precision as the foundations for rigid towers. With adjustable fittings, the structure can be plumbed to any desired level as discussed in Section 7.

6.5 CONNECTIONS

Guy connections to structures vary based upon structure types and materials. Individual components of these connections can be designed using established procedures of structural design or according to manufacturers' recommendations.

6.5.1 Guy Connections to Poles

Wood, concrete, and light duty steel poles almost exclusively use thru-bolting or banding for connections. Engineered steel poles or H-frames may have a variety of welded and bolted guy connections.

6.5.2 Connections in Latticed Structures

Critical latticed structure connections include the guy attachment points and masts' connections to foundations, guys, or bridge.

Most mast to footing connections are theoretically pinned even though the actual details of the connection often prevent it from rotating beyond a certain range. This is not a problem as masts do not rotate more than a few degrees during construction or during their lifetime. Footing connections for the guyed V tower must be able to accommodate the various angles the mast must make with the base for different height structures. The connections can vary from the very simple to complex. Figure 6-1(a) shows a connection made with a steel pin, embedded in concrete or directly into rock, a steel plate on top of a grouted pad, and a spherical washer. The base plate of the mast is allowed to deform slightly as it fits under pres-

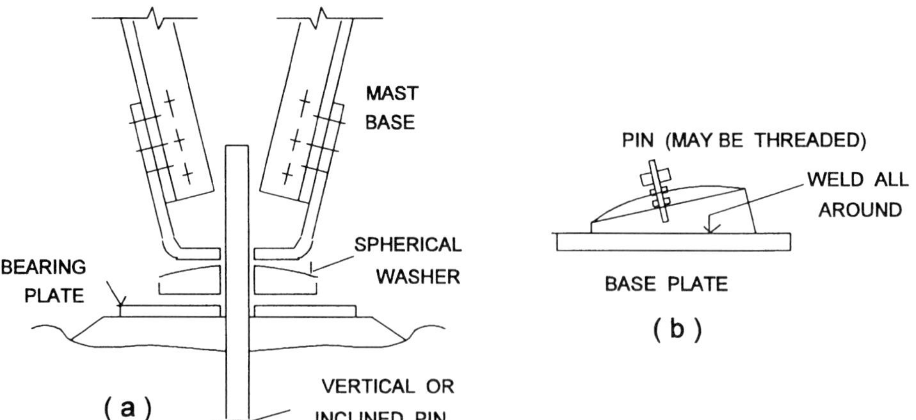

FIG. 6-1. Pinned Connection at Mast Base.

sure on top of the spherical washer. This one-time deformation is acceptable as long as it does not exceed a few millimeters. The spherical washer can be replaced by a welded assembly as shown for an inclined mast in Fig. 6-1(b). Ball and cup forgings without pins have been used (Fig. 6-2) but there is the danger that the base of the mast may slip out of the cup under unusual dynamic loading conditions. Dimensions of ball and cup forgings should avoid ''lock-up'' and proper drainage should be provided. Gimbal (universal joint) connections are also possible.

6.6 STRUCTURAL DESIGN OF POLES AND H-FRAMES

Individual components of guyed poles and H-frames should be designed using the same formulae and principles as similar unguyed structures. The main difference between guyed and unguyed poles and frames is that the possibility of buckling failure should always be checked in design in addition to the possibility of local failure from an excessive combination of bending, axial, or shear stresses.

Guyed wood poles in the past were traditionally analyzed assuming that the guys carried all the horizontal load, that is, by using the ''column'' assumption described in Section 5.3.1. Column buckling was evaluated with the formulae of Section 5.5.1. Since wood poles are relatively flexible and also tend to creep into the load with time, and since large safety factors against buckling are often used, the traditional design approach has usually proved adequate. Safety factors of 2 to 3 against buckling have been recommended (REA 1992) for unfactored NESC light, medium, and heavy loads. However, with the increasing availability of easy-to-use nonlinear analysis software on the microcomputer, there is no longer the need to treat wood poles differently from steel and concrete poles. With wood,

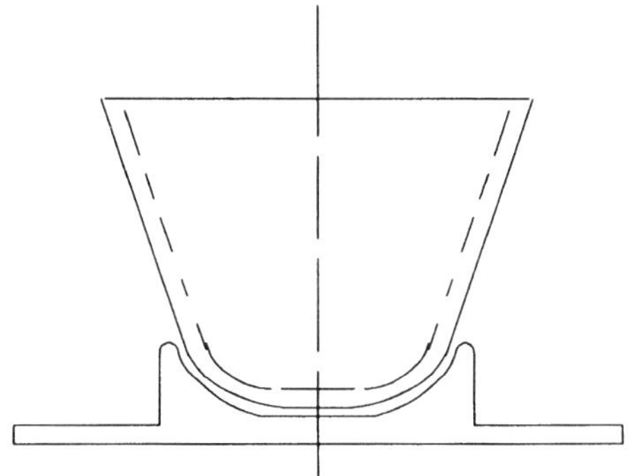

FIG. 6-2. Ball and Cup Forgings.

steel, and concrete treated identically, it should be possible to prepare comparable designs out of the three materials, that is, designs which produce poles with the same reliability.

For guyed steel and concrete poles, it has long been recognized that a computerized nonlinear analysis is necessary. Once the stress resultants (axial loads, shears, and moments) from the nonlinear analysis are known, the design procedures are the same as those used for an unguyed pole (ASCE 1990).

Guyed wood H-frames have traditionally posed even more problems than guyed wood poles since buckling and bending amplified by large axial forces cannot be neglected. A nonlinear computer analysis removes all questions and is thus the recommended procedure.

6.7 STRUCTURAL DESIGN OF LATTICED RIGID FRAMES AND MASTED TOWERS

Once the forces in the various members of a guyed latticed structure have been determined by one of the analysis methods presented in Section 5, preferably a geometrically nonlinear analysis, the design of individual members and connections is the same as for an unguyed structure (ASCE 1972; ANSI 1991).

The masted towers of Section 2.6 can be analyzed and designed manually with the procedures described in Section 5.5.2. A complete design is illustrated in Section 8. The manual procedure, or a spreadsheet type computer equivalent, is highly recommended at the preliminary design stage as it is the best way to understand how the structure functions and what its important design parameters are.

Chapter 7

CONSTRUCTION AND MAINTENANCE

The purpose of this section is to identify construction and maintenance practices that are necessary in order to ensure that the final installation of the guyed structure adheres to the assumptions made in design and to minimize actions, during the service life of the line, that could compromise the integrity of the structure.

7.1 DESIGN CONSIDERATIONS

The line designer should review and define limits for acceptable methods of construction and maintenance appropriate to the site conditions, applicable equipment, and skill level of the workers. Structural or other details that relate to the safety of construction and maintenance work should be considered in the design of the structure. The line designer should consult with constructors and maintenance personnel and utilize experience to develop a reasonable balance between construction flexibility and efficiency versus cost. Maximum anticipated construction and maintenance loads and their locations should be specified by the line designer.

7.2 CONSTRUCTION CONSIDERATIONS

Ease of construction can be enhanced by a number of considerations, both in the design of the structure and in detailing of the connections. It should be noted that these considerations could increase material costs, although these costs may be offset by reduced field costs and improved safety. Field assembly and erection methods can be influenced by the line design, line route, terrain, climatic or seasonal weather conditions, the impact of any environmental restrictions, line route access, schedule requirements, and the availability of critical resources in both manpower and equipment. For example, helicopter construction might be considered where movement of a large erection crane is difficult.

Guidelines on the erection of transmission structures are available (IEEE 1988). There are two aspects of a guyed structure construction that might influence the behavior of the final structure: the placement of the mast/pole footings and guy anchors, and the subsequent adjusting of the guy lengths and the tensioning of the guys. Major savings can be made in erection time and cost by quick-connect devices for guy terminations and by taking advantage of the relaxed tolerances permitted by the flexibility of some types of guyed structures. The tolerance limits should be based on the necessary and sufficient performance criteria of the installed structure.

7.3 GUYED POLES

7.3.1 Erection Methods

It should be noted that guyed poles and H-frames with fixed bases (anchor bolt/base plate type or direct embedded) will not have the relaxed erection tolerance of the guyed rigid or masted towers. The location accuracy of the guy anchors is important to the proper functioning of the guyed pole. The design angles of the guys in the vertical and horizontal planes in relation to the pole should be maintained and deviations from the guy configuration should be approved by the structure designer. Guyed poles may require two-point lifting or other special rigging to prevent excessive deflection and overstress during the lift.

A guyed steel pole with slip-type joints should be assembled on the ground with jacking devices. Variations in pole lengths caused by slip lap tolerance can be compensated through placement on the anchor bolts or through embedment length modifications. The jacking force should be at least equal to the maximum design load. Due to the need for a minimum tolerance in the pole length or where high axial force will result from guying, the structure designer may wish to consider using bolted flange connections as a substitute for slip joints.

7.3.2 Guy Installation

For guyed poles with fixed base, some designs may require immediate guy installation to resist a specified wind loading; other designs may use guys only to resist applied conductor and ground wire loads. The structure designer should identify to the installer the need for timely guy installation and any temporary guying required to provide structure stability prior to the line completion.

For angle or dead-end poles or H-frames, the guys are usually installed so that the structure is plumb or has a slight negative deflection under the transverse or longitudinal pulls from everyday cable tensions. This can be done in several ways: (1) by adjusting the guy tensions after the line has been strung, (2) by pretensioning the guys against the pole prior to stringing in such a way that the top of the pole has a specified amount of negative deflection, or (3) by pretensioning the guys against the pole prior to stringing at specified tensions or at a "snug tight" or "taut" condition. "Snug tight" or "taut" are unfortunately vague terms which refer to the removal of all slack in the guy.

For tangent poles or H-frames with pairs of transverse guys and/or pairs of longitudinal guys (head and back), the guys in each pair are usually installed "snug tight."

If guy pretensions are specified, they should be applied as specified by the structural engineer within the construction tolerances outlined in Section 7.5.1.

Specific recommendations for direct-embedded poles are also required from the foundation designer as to the type of material and method of placement of backfill in order to provide the anchorage necessary to develop the strength of the pole without excessive rotation in the soil. Following these procedures it is necessary to ensure that the behavior of the pole in service is in accordance with the design assumptions.

7.4 GUYED RIGID FRAMES AND MASTED TOWERS

7.4.1 Erection Methods

Guyed frames and towers may be erected by any method that suits the terrain and access conditions, worker experience, and available equipment.

Vertical loads have negligible effect on the guy tensions of tangent structures so that adjustment of guy tensions after stringing is seldom, if ever, required. There may be a need to adjust guy tensions after stringing through a guyed angle structure.

7.4.2 Crane Erection

When access is suitable, the use of a crane has proven to be a useful and economic method of erection. The necessary clearing of the site to permit assembly of the structure and the positioning of the crane requires some planning.

The guyed structure should, if possible, be tilted up as a complete unit, although, at particularly difficult sites, guyed structures have been erected section by section using temporary guys (e.g., with a gin pole).

7.4.3 Helicopter Erection

The availability of helicopters with increased lift capacities combined with the reduced weight of guyed structures has resulted in an increase in helicopter-based construction. Increased environmental pressures limiting road access and on-site clearing have added to the situation.

Helicopters can be used to lift the entire structure (sometimes complete with insulator assemblies, stringing blocks, and finger lines) from the assembly yard to the site, or to tilt-up the structure assembled on the ground directly on its foundation.

Moving the assembly work from the tower site to a flat open assembly yard has the advantages of:

1. reducing the loss of pieces and elimination of delay due to such losses;
2. reducing clearing and grading of the site, which can be a major problem when trying to produce a large flat area on which to lay out a large structure;
3. reducing the costs of transporting the assembly workers to the various sites;
4. reducing assembly costs. Assembling as many as 40 or more structures at one site produces economies when power equipment can be used for bolt tightening, jigs set up for assembling the masts, and the guys can be measured, cut, fitted, and attached before the structures are moved; and
5. reducing environmental impact (minimum access road work).

Other items to consider when planning helicopter erection are:

1. provisions for attachment of guide members to the structures;
2. adequacy of helicopter lifting capacity at the site altitude and ambient temperature;
3. proper consideration of weights, centroids, and lifting points; and
4. whether workers will be allowed on or under a structure during setting.

The greatest efficiency in savings of cost, time, materials, and impact on the land will be reached when the helicopter is introduced at the start of the design work and planning of construction.

7.4.4 Guy Installation

The top ends of the guys have a permanent nonadjustable dead-end fitting and these are attached to the top of the masts or crossarms before erection of the structure. The remainder of the operation of attaching the guys to the anchors, the plumbing of the structure, and then the pretensioning of the whole system can be done by at least two different methods.

7.4.4.1 Traditional Method. The guys are cut to approximate length (but always slightly longer than needed), permanently attached at their tops to the structure, and attached at their bottoms with temporary grips or other wedge-type devices. The structure is held in approximate plumb position by crane or helicopter while some block and tackle or tensioning device takes up the slack in each guy and holds the structure in place.

Surveying equipment and plumb bobs are used while the structure is brought into proper alignment by tightening and slackening the guys; the loads are then transferred to the permanent fitting which contains threaded tensioning devices used to apply the specified pretension.

This operation can involve many person-hours, especially if close and rigid specifications have been set regarding the positioning of the tower and the pretension levels.

7.4.4.2 Alternate Method. This method is advantageous if one accepts the fact that relaxed or more liberal than usual tolerances can be applied to the erection of guyed rigid frames and masted towers. These relaxed tolerances often will have negligible effect on the strength or performance of the structures, the tolerances on erection, and positioning of the structure being limited to what is needed to ensure that the structure does not appear other than vertical.

The human eye cannot detect if a standalone (not in a line) mast or tower is out of plumb by less than about one degree. One-half of a degree has been accepted by some as a limit that would permit the center of the crossarm to be no more than 1% of its height out of place. Furthermore, with tangent suspension towers such as the guyed V, guyed Portal, guyed Delta, or the CRS, the tensions in all four guys, in the absence of wind, are all equal and any pretensioning can be introduced by using a threaded tensioning device in but one of the guys.

Acceptance of a limit of one-half degree in verticality and applying tension to one guy permits a great reduction in the time required to position a tower. It also allows reduction in the number of tensioning devices.

A survey is made of the elevations of the tops of the anchors and of the mast footing(s) and calculations made of the needed guy lengths. Guys are then cut and three of them fixed to length with simple dead-end fittings; an adjustable device added to one only. Upon erection of the structure, it is moved slightly at the top to permit direct attachment of three of the guys to their anchors and finally the fourth is attached and the desired pretension introduced.

The alternate method requires that the lengths of the three guys and their fitting be accurate to about 0.3% of calculated length, which is a reasonable degree of precision.

7.4.4.3 Guy Pretension. Guy pretensions for rigid frames and masted towers are discussed in Section 6.3.1. As all four guy tensions for typical tangent guyed Vs, Portals, and Deltas are essentially equal, there is no need for elaborate procedures to check all the guy tensions.

7.5 ERECTION TOLERANCES

The erection tolerances have to be specified for the parameters:

1. the vertical alignment of the tower;
2. the anchor locations, the anchor rod alignment, and the backfill compaction; and
3. the precut length of fitted guys.

7.5.1 Guyed Poles and H-Frames

Guyed poles should be set in a straight vertical position. The horizontal offset tolerance should be within 75 mm (3 in.). The vertical tolerance should be within 150 mm (6 in.) from the specified setting depth.

On guyed H-frames, a maximum of 75 mm (3 in.) elevation difference is allowed between the two adjacent poles and a maximum of 50 mm (2 in.) horizontal offset between the two H-frame legs. Guy anchors can be relocated within 1 m (3 ft) radius of the original design location and the repositioning of the anchor should be documented and reported to the engineer for further analysis. A tolerance of 10% of the specified guy pretension is considered practical.

7.5.2 Guyed Rigid Frames and Masted Towers

The overall verticality of the structure should be based on a limiting requirement that the structure does not appear to be out of plumb to the human eye. A recommended value is one-half degree. Thus, for a structure such as the guyed V or the guyed Portal, the center point of the crossarm should not be more than one-hundredth of the height from the structure center point on the ground.

The anchor is a foundation in uplift and normal foundation construction tolerances should be applied. The main difference is the location flexibility. Relocating the anchor may cause the guy to load the structure differently from that assumed in design. These changes in structural loading need to be considered before construction begins. The location of the anchor rod should have a tolerance of ±2% of the structure height from the staked location. A check should be carried out for possible electrical conflicts. The orientation of the anchor with respect to the guy cable may have to be changed. The anchor rod alignment should be within ±5 degrees to the specified guy angle. For backfilled ground anchors, backfill compaction should be expressed in terms of relative density for granular material and Proctor density for fine-grained noncohesive and cohesive soil. The quality control of the backfill is critical.

A recommended value of ±10% tolerance limit of the specified initial guy tension should be used.

7.6 INSPECTION AND MAINTENANCE

Like any other transmission structures, guyed structures should be given a detailed inspection by a professional engineer or line worker at periodic intervals. The inspector should submit a written report with recommendations. For guyed structures, the inspection should include the following items in addition to what is normally required for an unguyed structure.

1. Alignment: tolerances should not exceed the erection tolerances.
2. Guy tensions should be within erection tolerances of the specified construction values.
3. Slack guy assemblies: a slack guy on a pole or a set of slack guys on a guyed structure usually indicate foundation or anchor movement. In any case, slack assemblies should be checked for:
   a. broken guy strands,
   b. slippage of guy grips, clips, and the like,
   c. loose, worn, cracked, bent, or missing hardware,
   d. articulation at guy ends and, in particular, at turnbuckles, and
   e. turnbuckle take-up.

Chapter 8

EXAMPLES

This section includes several examples that illustrate many of the concepts presented in this guide. The examples and load cases have been simplified and may not represent actual design conditions.

8.1 WOOD POLES

The ultimate buckling capacity of a wood pole with two different guying systems (Fig. 8-1) is determined with manual methods and compared with what can be obtained with a nonlinear computer program. It is shown that errors in excess of 100% on either the conservative or nonconservative side are possible with manual methods.

8.1.1 Dead-End Pole with In-Line Guys in Single Vertical Plane

All four guys have a slope of 45 degrees and are attached at the locations of the ground-wire loads (HT and VT) and conductor loads (HC and VC at each of three conductor locations). There is no guying out of the vertical plane containing the ground wire and the three conductors (Fig. 8-1(a)).

The guyed wood pole properties are as follows.

    Guys (at 45 degrees)
        $A \times E$ = 19,600 kN   (4,400 kips)
        Final tension   = T1, T2, T3, T4 kN   (kips)
    Pole
        Distance a = 3.05 m   (10 ft)
        Distance b = 14.0 m   (46 ft)
        Top diam. DHT = 21.8 cm   (8.6 in.)

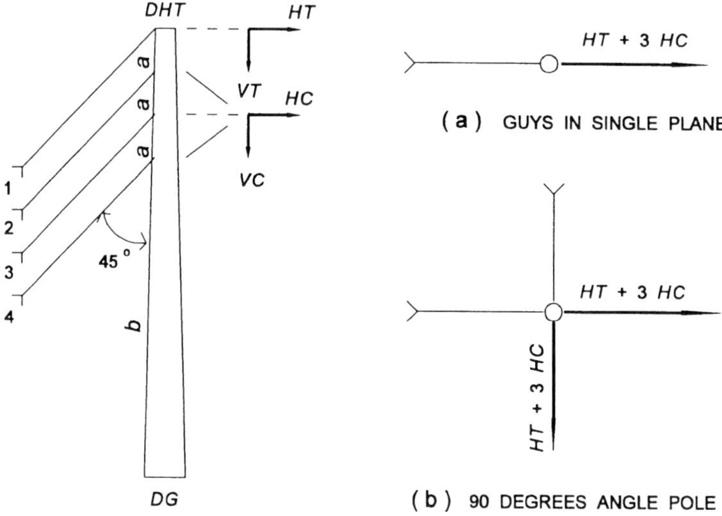

FIG. 8-1. *Wood Pole Example.*

Diam. at guy 3	= 27.4 cm	(10.8 in.)
Diam. at guy 4	= 30.2 cm	(11.9 in.)
Ground diam. DG	= 42.9 cm	(16.9 in.)
E	= 13,100 MPa	(1,900 ksi)
Ultimate bending stress	= 55.1 MPa	(8 ksi)
Base compression	= P kN	(kips)
Buckling cap.	= P_{CR} kN	(kips)
Basic loads		
HT	= 22.2 kN	(5 kips)
VT	= 4.45 kN	(1 kip)
HC	= 44.5 kN	(10 kips)
VC	= 8.90 kN	(2 kips)
Pole weight	= 8.90 kN	(2 kips)

8.1.1.1 Analysis and Buckling Capacity by Manual Methods. Using the "column" or "strut" method described in Section 5.3.2, the guy tensions and base compression are simply: $T1 = 31.6$ kN (7.1 kips), $T2 = T3 = T4 = 62.7$ kN (14.1 kips), and $P = 5 + 1 + 3 \times (10 + 2) + 2 = 196$ kN (44 kips).

With in-line guys and no lateral support, the buckling shape is assumed to resemble that shown in Fig. 5-11(d) for a pole with a single in-line guy; that is, the equivalent unbraced length factor K is equal to 1. Because the example pole has four guys that introduce compression loads at different locations, the effective pole length L has to be assumed. Buckling loads, based on different assumptions, have been calculated as shown in Table 8-1. The first three lines show results obtained

TABLE 8-1 Buckling Load Calculations for Various Assumptions: Tangent Dead-End Wood Pole

| Effec. Length L | | Top Diameter | | Moment Inertia | | P^* in | P_{CR} | |
m	(ft)	cm	(in.)	cm^4	(in.^{4})	Eq. (5-7)	kN	(kips)
23.2	(76)	21.8	(8.6	11,150	(268)	3.86	104	(23.3)
17.1	(56)	27.4	(10.8)	27,800	(668)	2.45	302	(67.9)
14.0	(46)	30.2	(11.9)	40,960	(984)	2.02	543	(122)
14.0	(46)	33.0	(13.0)	58,350	(1402)	Not used	384	(86.3)

by the Gere and Carter method (Eq. (5-7) in Section 5.5.1), using different length assumptions (i.e., assuming that a single guy load is applied at the top of a pole of length L). The buckling load in the last line was calculated with Eq. (5-6) using a uniform moment of inertia equal to that at the section of the pole at 7/9 of the height to the lowest guy attachment point (REA 1992).

The results in Table 8-1 show that there can be considerable variation in the manual estimate of the buckling load P_{CR}, depending on the assumptions made.

8.1.1.2 Analysis and Buckling Capacity by Nonlinear Computer Analysis. A geometrically nonlinear finite element model of the pole was generated by discretizing it with prismatic elements not exceeding about 1.5 m (5 ft) in length. The guys were modeled with true cable elements even though straight bars would have been perfectly appropriate. For the basic loads described in Section 8.1.1, the computer analysis yielded the following values: $T1 = 36.2$ kN (8.13 kips), $T2 = 56.5$ kN (12.7 kips), $T3 = 64.5$ kN (14.5 kips), $T4 = 62.1$ kN (14.0 kips), and $P = 204$ kN (45.9 kips). The guy forces and base reaction for the nonlinear analysis are essentially the same as those obtained by the manual method.

With a nonlinear computer program, the ultimate capacity of a structure is determined by increasing the basic loads by a constant load factor LF until either some stress reaches its ultimate value or the solution does not converge, an indication that some instability has occurred. In order to trigger instabilities in a perfectly symmetrical computer model, it is sometimes necessary to break the symmetry. In the program which was used, this is done automatically by the application of small transverse and longitudinal wind pressures on the pole when no pressure is specified in the design.

Table 8-2 shows the behavior of the pole as the load factor is increased. X represents the displacement at the top of the pole in the direction perpendicular to the vertical plane of the guys. The results in Table 8-2 clearly show that the pole becomes unstable for a load factor of about 1.05 which corresponds to a vertical base reaction (buckling

TABLE 8-2 Guyed Wood Pole Behavior Under Increasing Loads: Tangent Dead-End Pole

Load Factor	Base Reaction		X-Displacement		Maximum Stress	
	kN	(kips)	cm	(in.)	kPa	(psi)
0.9	186	(41.8)	3.3	(1.3)	—	
1.0	204	(45.9)	17	(6.6)	9,200	(1,330)
1.05	210	(47.3)	586	(231)	81,800	(11,870)
1.10 (Max)	213	(47.8)	935	(368)	136,000	(18,040)

load P_{CR}) of 210 kN (47.3 kips). The computer-based buckling value of 210 kN can be compared to the manual estimates in Table 8-1.

This example demonstrates that, depending upon the assumptions made, the manual method can produce buckling values from one-half to two and one-half times the more reliable values obtained by computer. Similar conclusions were reported in a paper by Peabody and Wekezer (1994).

8.1.2 Ninety-Degree Angle Pole with In-Line Guys

The pole in this example is identical to that in Section 8.1.1, except that the tension loads are now coming from two directions, thus requiring two sets of guys as shown in Fig. 8-1(b).

8.1.2.1 Analysis and Buckling Capacity by Manual Methods. Using the ''column'' method described in Section 5.3.2, the guy tensions are the same as those in Section 8.1.1.1. The base compression P is equal to $2 \times (22.2 + 4.45 + 3 \times (44.5 + 8.90)) + 8.90 = 383$ kN (86 kips).

With two sets of in-line guys providing some lateral support at the guy levels, the buckling shape is assumed to resemble that shown in Fig. 5-11(c) (fixed, pinned); that is, the equivalent unbraced length factor K is equal to 0.7. Because the example pole has eight guys that introduce compression loads at different locations, the effective pole length L has to be assumed. Buckling loads, based on two different assumptions, have been calculated as shown in Table 8-3. The first line shows results obtained by the Gere and Carter method (Eq. (5-7) in Section 5.5.1), using the common assumption that all vertical loads are lumped at the attachment point of the lowest guy. The buckling load in the last line was calculated with Eq. (5-6) using a uniform moment of inertia equal to that at 7/9 of the height to the lowest guy attachment point (REA 1992).

8.1.2.2 Analysis and Buckling Capacity by Nonlinear Computer Analysis. A geometrically nonlinear finite element model similar to that described in Section 8.1.1.1 was used to generate the data in Table 8-4. The displacement X represents the lateral displacement at the top of the pole

TABLE 8-3 Buckling Load Calculations for Various Assumptions: Angle Wood Pole

Effec. Length L		Top Diameter		Moment Inertia		P^* in	P_{CR}	
m	(ft)	cm	(in.)	cm^4	(in.^{4})	Eq. (5-7)	kN	(kips)
14.0	(46)	30.2	(11.9)	40,960	(984)	2.02	1,112	(250)
14.0	(46)	33.0	(13.0)	58,350	(1402)	Not used	783	(176)

in the direction of the conductors. Actually, larger displacements occur in the direction of the line angle bisector.

The results in Table 8-4 show that the pole fails in bending for a load factor of about 2.85 when the bending stress approaches the ultimate 55,200 kPa value (8,000 psi). For load factors above 2.85, the pole displacements in the direction of the line angle bisector and the stresses increase rapidly, indicating impending buckling.

For this example the computer value of 1,097 kN (247 kips) is near the Gere and Carter value in Table 8-3 (1,112 kN, 250 kips), but exceeds the REA value (783 kN, 176 kips). The buckling shape is similar to that shown in Fig. 5-12, with a point of contraflexure near the lowest guy attachment point.

8.2 TUBULAR STEEL POLES

The conclusions drawn from the examples in Section 8.1 for wood poles are also valid for tubular steel poles. Caution must be exercised when using manual methods of analysis as they can produce very misleading results.

TABLE 8-4 Guyed Wood Pole Behavior Under Increasing Loads: Angle Pole

Load Factor	Base Reaction		X-Displacement		Maximum Stress	
	kN	(kips)	cm	(in.)	kPa	(psi)
1.0	391	(88)	7	(2.8)	14,200	(2,060)
2.85	1,097	(247)	25	(9.7)	53,900	(7,820)
3.0	1,158	(260)	27	(10.6)	71,800	(10,410)
3.1	1,199	(270)	30	(11.7)	93,800	(13,600)
3.15	1,120	(275)	33	(12.9)	114,000	(16,600)

The example in Section 8.2.1 compares the results of a pole designed using a linear ''column'' analysis method with that of one designed using a computer-aided nonlinear technique. This example demonstrates why nonlinear methods are recommended for all guyed steel pole analyses.

Design can also be affected by construction techniques and tolerances. The examples in Section 8.2.2 show the effects and significance of altering basic guy geometry as might be necessitated by actual field conditions.

8.2.1 Bisector Guyed Pole

Consider the guyed steel pole of Fig. 8-2 which has the following properties.

Guys (at 45 degrees)   Extra High Strength   (1 × 19)
   Diameter   = 14.3 mm   (9/16 in.)
   Max. tension   = 97.4 kN   (21,900 lbs) 65% of RBS
   $A \times E$ = 19571 kN   (4,400 kips)
Pole (12-sided cross-section shape)
   Distance a = 3.05 m   (10 ft)
   Distance b = 21.3 m   (70 ft)
   Top diam. DHT = 22.9 cm   (9.0 in.)   (flat to flat)
(1) Ground diam. DG = 30.5 cm   (12 in.)   size adequate by
                                                    ''strut'' approach
(2) Ground diam. DG = 62.2 cm   (24.5 in.)   size required by
                                                    design stresses with
                                                    nonlinear analysis
                                                    approach
   Plate thickness   = 4.76 mm   (.1875 in.)
Basic loads (NESC Heavy)
   HT = 32.60 kN   (7.33 kips)
   VT = 5.38 kN   (1.21 kip)
   HC = 63.38 kN   (14.25 kips)
   VC = 15.74 kN   (3.54 kips)

Using the ''column'' or ''strut'' method described in Section 5.3.2, the compression load P in the pole (ignoring its weight) is equal to $5.38 + 32.60 + 3(15.74 + 63.38) = 275.3$ kN (61.9 kips) and there is no moment at the base and no deflection at the top.

Using the Gere and Carter method (Gere and Carter 1962) and assuming an effective pole length of 21.34 m (70 ft) with a K factor of .7 (i.e., lateral support at the lower guy), the buckling capacity P_{CR} is calculated as follows.

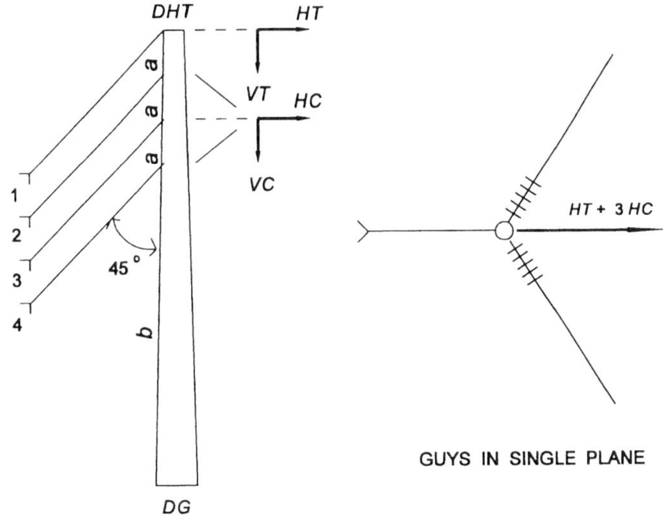

FIG. 8-2. Steel Pole Example.

Ratio of base diameter to diameter at lower guy: d_g/d_a = 30.5/ 25.1 = 1.21

Ratio of base to top moments of inertia: I_g/I_a = 5300/2950 = 1.80

Shape factor $n = \log [I_g/I_a]/\log [d_g/d_a]$ = 3.08

P^*(From Fig. 5 of Gere and Carter paper) = 1.50

$P_{CR} = P^* \pi^2 E_S I_a/(K\,L)^2 = 1.5\,(3.1416)^2\,(200 \times 10^6 \times 2950 \times 10^{-8}/ (0.7 \times 21.34)^2 = 391$ kN (87.9 kips) > 275.3 kN (61.9 kips)

The preceding calculations suggest that the pole with a ground diameter of 30.5 cm (12 in.) and lateral support at the location of the lower guy is satisfactory. However, if one assumes that there is no lateral support provided by the conductors at the lowest guy attachment point (i.e., K = 1 or larger), the pole buckling strength is inadequate. In addition, with the linear "column" or "strut" approach, the guys are assumed infinitely stiff. In reality, when loaded, the guys stretch allowing the pole to deflect. For a pole with any degree of fixity at its base there is both an axial force and moment at the base. There are also forces and moments along the shaft. Only by using a nonlinear computer program can these forces and moments be easily and accurately calculated and the pole sized for the actual stresses.

For this example, a nonlinear analysis, which assumes that there is no restraint provided by the ground wires and conductors, shows that a substantially larger pole (with a base diameter of 62.2 cm and 56% heavier) is needed over what the "column" or "strut" method requires. This is shown in Fig. 8-3 where both the "strut" design and the more accurate nonlinear design are contrasted. Both poles have the same plate thickness, 4.76 mm (.1875 in.). The actual tensions in the guys are substantially different from those based on the linear "strut" method.

8.2.2 Effect of Guy Properties on Behavior of Pole

This section describes the effects of changing any one of the geometric guy parameters of a tubular steel pole. The base structure is the pole described in Section 8.2.1 with a groundline diameter of 62.2 cm (24.5 in.).

The following perturbations have been imposed on the basic design.

1. Guy slopes increased from 1 : 1 to 1 : 2 (vertical : horizontal)
2. Base fixity lowered from ground line by one-third of the embedment length (lowered 1.52 m)
3. Guy lengths increased by factor of 1.5
4. Guy sizes decreased from 14.3 mm (9/16 in.) to 12.7 mm (1/2 in.)

Table 8-5 summarizes the effects of the changes.

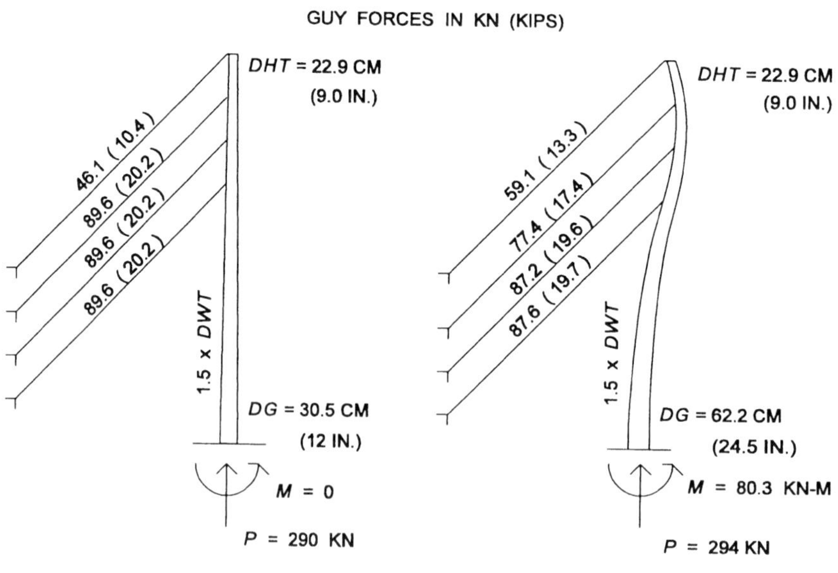

FIG. 8-3. "Strut" Versus Nonlinear Analysis and Design of Steel Pole.

TABLE 8-5 Effects of Changing Guy Parameters on Response of Steel Pole

	Guy Forces				GL Reactions			Top Defl.
	SW	Top Con	Mid Con	Low Con	Axial	Shear	Moment	
Base design (in Fig. 8-2)	1.	1.	1.	1.	1.	1.	1.	1.
Guy slope changed to 1 : 2	1.49	1.42	1.38	1.35	1.49	1.26	1.40	1.75
Fixity at 1/3 embedment depth	1.	1.	1.	1.02	1.01	0.97	0.99	1.
Guy length increased 50%	1.26	1.01	0.97	0.92	0.99	1.37	1.50	1.57
Smaller guys	1.03	1.00	0.97	0.98	1.00	1.15	1.19	1.30

8.3 GUYED V

Figure 8-4 shows the dimensions (in meters) and loads for a guyed V tower. Hand calculations are shown for the guy and mast forces and the preliminary design of the masts.

8.3.1 Analysis for High Wind Loads

The guy and mast forces are to be determined for: (a) conductor loads $VC = 40$ kN and $TC = 58$ kN; (b) transverse wind load on the crossarm $TA = 5$ kN; and (c) a transverse wind pressure of 0.6 kN/m on the masts. Guys 1 and 2 will share the wind load equally, and guys 3 and 4 will go slack.

Overturning moment
about base A:

$$
\begin{array}{lr}
3 \times 58 \times 34 & = 5{,}916 \text{ kN} \\
5 \times 35 \text{ (est.)} & = \quad 175 \\
2 \times 0.6 \times 34 \times 34/2 & = \quad 694 \\
\hline
\text{OTM} & = 6{,}785 \text{ kN}
\end{array}
$$

Vertical load at guy
anchor $\qquad = 6{,}785/2 \times 22 \qquad = 154$ kN

Guy tension $\qquad = 154 \times 40.5/34 \qquad = 183$ kN

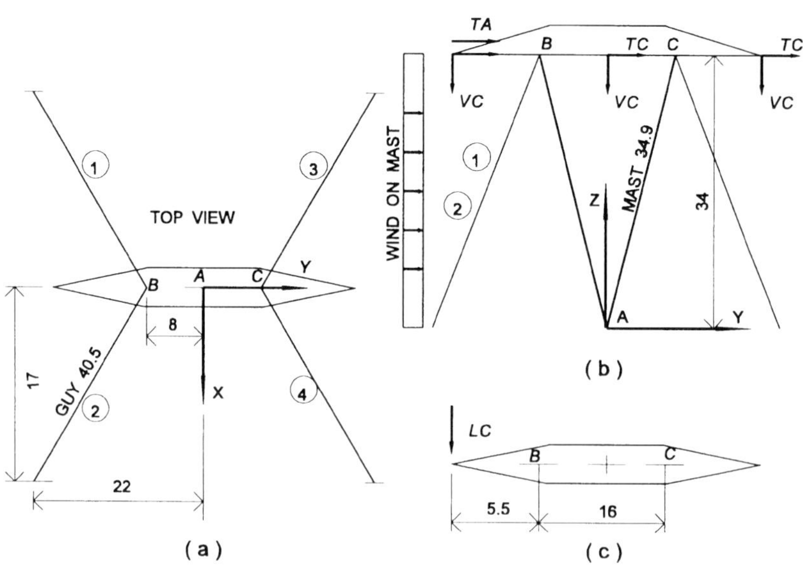

FIG. 8-4. *Overall Configuration of Guyed V Example.*

RBS required for
  design limit
  of 70%: $RBS = 183/.7$ $= 261$ kN
Vertical windward
  mast load: from two guys $= 2 \times 154$ $= 308$ kN
                     from conductors $= 3 \times 40/2$ $= 60$
                     from tower weight
                       (est.)/2 $= 10$
                                                           ____
                                                            $= 378$ kN

Mast compression
  load $= 378 \times 34.9/34$ $= 388$ kN

8.3.2 Analysis for Unbalanced Longitudinal Load on Outer Phase

The guy and mast forces are to be determined for the torque load produced by a single longitudinal load $LC = 65$ kN at the outer phase (see Fig. 8-4(c)). Guy 2 will go slack. The longitudinal component of the tension in Guy 1, $G1X$ is calculated by writing a moment equation about the vertical axis going through point C:

$$G1X = 65 \times (16 + 5.5)/16 \qquad = 87.5 \text{ kN}$$
$$\text{Guy tension} = 87.5 \times 40.5/17 \qquad = 208 \text{ kN}$$
$$\text{RBS required for design limit of } 85\% \qquad = 208/.85 \quad = 244 \text{ kN}$$

(higher limit than in Section 8.3.1 because emergency failure containment load, or security load, is involved).

RBS required of 244 kN is less than 261 kN required in Section 8.3.1. Therefore, for the loads of Sections 8.3.1 and 8.3.2, select guy with RTS larger than 261 kN such as 19mm EHS with a RBS of 311 kN.

8.3.3 Analysis for Combination of Vertical, Transverse, and Longitudinal Loads

The guy and mast forces are to be determined for a combination of conductor loads VC, TC, and LC, transverse wind load on the crossarm TA, and transverse wind pressure on the masts. Assuming that LC is large enough, Guy 2 will go slack. Otherwise, Guy 3 will go slack. The longitudinal components of guy forces are $G1X$, $G2X$, $G3X$, and $G4X$. Their vertical components are $G1Z$, $G2Z$, $G3Z$, and $G4Z$. All of these components are directly related to the guy tensions $G1$, $G2$, $G3$, and $G4$. These four unknown guy tensions, which must be positive or zero, can be determined by using the following four equations.

1. One guy is slack $G2$ or $G3 = 0$
2. X-axis overturning $22 \times (G1Z + G2Z - G3\,Z - G4Z) = OTM$
   about A
3. Equilibrium in long. $G1X + G3X - G2X - G4X = LC$
   direction
4. Z-axis moment
   about C (if $G2 = 0$) $16 \times (G1X) = 21.5 \times LC$
   or moment about B
   (if $G3 = 0$) $16 \times (G4X) = 5.5 \times LC$

Once the guy tensions are known the mast forces can easily be established.

8.3.4 Design of Mast for Guyed V

The mast for the guyed V of Fig. 8-4 will be designed as a square 1.1 m × 1.1 m latticed box with staggered bracing at 45 degrees as shown in Fig. 8-5. The maximum design compression load is 388 kN.

Overall slenderness ratio $L/r = 34.9/.55$ $= 64$ (approx.)
Mast moment from wind $= 0.6 \times 34^2/8$ $= 86.7$ kN-m
Mast P-Δ moment (assume $\Delta =$
   0.2m) $= 388 \times .2$ $= 77.6$ kN-m
P-Δ moment will be re-checked
   later
Chord load: from compression $= 388/4$ $= 97$ kN
            from wind moment $= 86.7/(2 \times 1.1)$ $= 39.4$ kN
            from P-Δ moment $= 77.6/(2 \times 1.1)$ $= 35.3$ kN

      Total $= 172$ kN

Chord member: Try 75 × 75 × 6 ($A = 8.64$ cm^2, $r_x = 23.3$ mm) with
   $F_y = 345$ MPa
Slenderness $L/r_x = 2{,}200/23.3 = 94.4 < C_c = 106$
Design compression stress $F_c = (1 - .5 \times 94.4^2/106^2) \times 345 =$
   208 MPa
Compression capacity $= 8.64 \times 10^{-4} \times 208 \times 10^3 = 180$ kN $>$
   Chord load $= 172$ kN (95% use) OK
Mast shear from wind $= 0.6 \times (34/2) = 10.2$ kN
Bracing member load $= 10.2/(2 \times .707) = 7.2$ kN

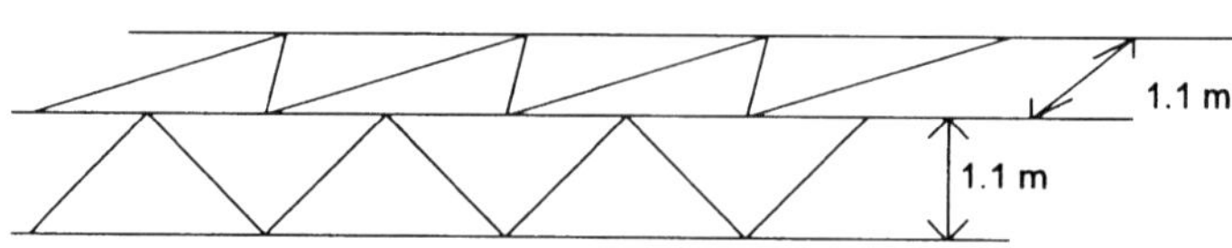

FIG. 8-5. Mast for Guyed V of Fig. 8-4.

Bracing member: Try $45 \times 45 \times 3$ ($A = 2.71$ cm^2, $r_z = 8.86$ mm) with
  $E = 200,000$ MPa
Slenderness $L/r_z = 1,100 \times 1.414/8.86 = 175 > C_c = 106$
Design compression stress $F_c = \pi^2 \times E/(L/r_z)^2 = 64$ MPa
Compression capacity $= 2.71 \times 10^{-4} \times 64 \times 10^3 = 17.3$ kN $>$
  Bracing load $= 7.2$ kN OK

Re-check of assumed P-Δ moment
Mast moment of inertia: $I = 4 \times 8.64 \times 10^{-4} \times .55^2 = 10.5 \times 10^{-4}$ m^4
Mast buckling load (Eq. (5-9)) $P_{CR} = .95 \times \pi^2 \times 200,000 \times 10^6 \times 10.5 \times 10^{-4}/ 34.9^2 = 1,617$ kN
Amplification factor (Eq. (5-8)) $AMPF = 1/(1 - 388/1,617) = 1.32$
Mast deflection from wind
  $\Delta_{wind} = 5 \times .6 \times 10^3 \times 34.9^4/(384 \times 200,000 \times 10^6 \times 10.5 \times 10^{-4}) = .055$ m

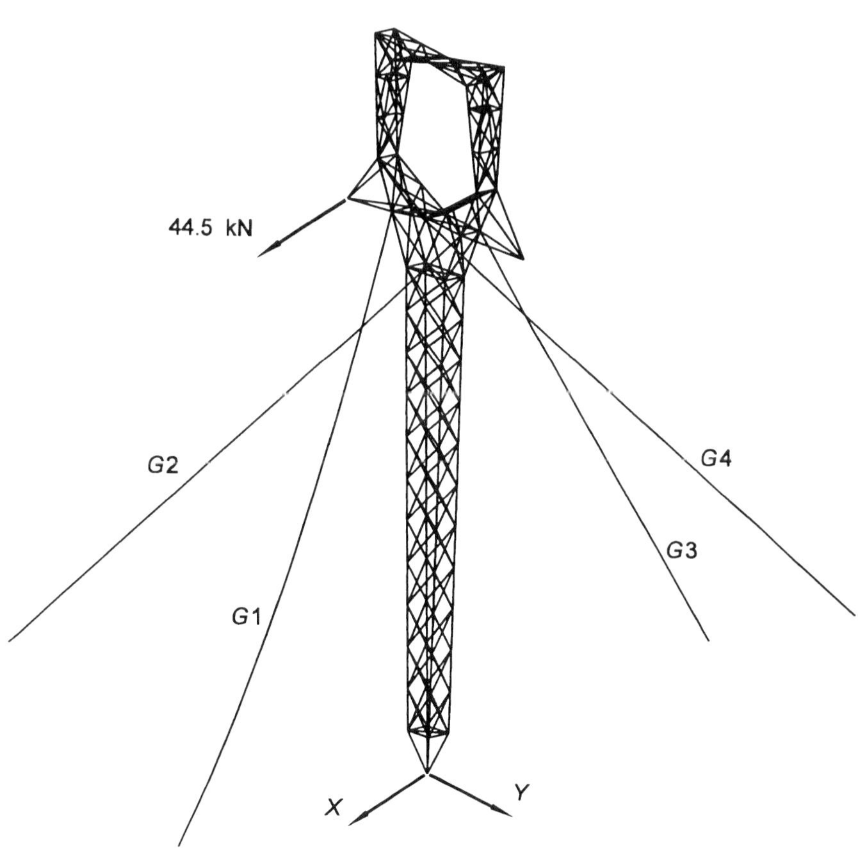

FIG. 8-6. Guyed Delta Example.

Additional mast deflection from out-of-straightness (see Section 5.5.2)

$$\Delta_{\text{out-of-straightness}} = L/300 = 34.9/300 = .116 \text{ m}$$

Total deflection including P-Δ effect $= (.055 + .116) \times AMPF$
$= .171 \times 1.32 = .23 \text{ m}$

Total deflection of .23 m close to assumed .20 m. Chord use was 95%—Design OK

8.4 GUYED DELTA

Consider the guyed delta tower in Fig. 8-6. The guy tensions will be determined for a broken outer phase longitudinal load of 44.5 kN (10 kips). The tower can be viewed as a large solid pinned at its base and supported by the four guys. Fig. 8-7 shows the important dimensions of the tower and its guys. The guys are attached to the structure at the apexes of rectangle $ABCD$ as shown in Fig. 8-7. The 44.5 kN load is equivalent to a torque of $44.5 \times 4.88 = 217$ kN-m applied at the center of the rectangle $ABCD$ together with a longitudinal load of $44.5 \times 19.81/20.08 = 43.9$ kN. The ratio 19.81/20.08 accounts for the fact that the actual longitudinal load is located slightly below the rectangle $ABCD$. It is clear that Guy No. 1 will go slack. There are therefore three unknown guy tensions to be solved. Their longitudinal components, labeled $G2X$, $G3X$, and $G4X$, are used as unknowns in the following equations.

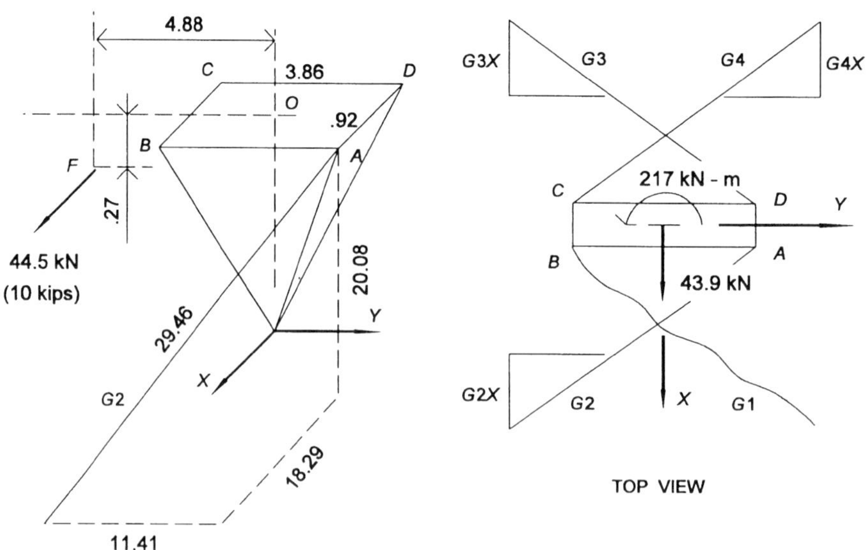

FIG. 8-7. Statics of Guys in Tower of Fig. 8-6.

TABLE 8-6 Final Guy Forces (in kN) in Tower of
Fig. 8-6 by Various Methods

	GUY 1	GUY 2	GUY 3	GUY 4
Manual analysis of this section	0	79	35	114
Computer analysis— no vertical load	1	82	36	116
Computer analysis— with vertical loads	1	86	40	117

For equilibrium of the rectangle $ABCD$ (see Fig. 8-7):

1. In Y direction $\quad(11.41/18.29)\,(-G2X - G3X + G4X) \qquad = 0$
2. In X direction $\quad -G2X + G3X + G4X \qquad\qquad\qquad = 43.9$
3. Mom. about $O\quad 3.86/2\,(G2X - G3X + G4X) + .92/2$
$\qquad\qquad\qquad (11.41/18.29)\,(G2X - G3X + G4X) \qquad = 217$

The three equations are solved simultaneously and the values of
$G2X$, $G3X$, and $G4X$ are multiplied by 29.46 / 18.29 to get the final
guy forces. These guy forces are summarized in Table 8-6 together
with those from two nonlinear computer solutions. Guy pretensions
of 25.9 kN and exact cable elements were used in the computer
solutions. The first computer solution was obtained without vertical
loads to match the assumptions of the manual solution. The second
solution was obtained with vertical loads of 4.45 kN (1 kip) from the
ground wires, 17.8 kN (4 kips) from the unbroken conductors, and
8.9 kN (2 kips) from the broken conductor. The computer analysis
without vertical loads showed a longitudinal displacement of 29 cm
(.94 ft) at the point of application of the longitudinal load. It can be
seen from the results in Table 8-6 that guy tensions obtained by
manual and computer methods are quite close with vertical loads
excluded from the analysis. When the vertical loads are included,
the differences are more significant.

Appendix A

REFERENCES

AISI (1981). *Wire rope user's manual.* 2nd edition, American Iron and Steel Institute, Washington DC.

ANSI/ASCE (1991). *Design of latticed steel transmission towers.* ANSI/ASCE Standard 10-90, New York.

ASCE (1972). *Guide for the design of aluminum transmission towers.* *J. Structural Division,* ASCE, December.

ASCE (1990). *Design of steel transmission pole structures.* ASCE Manual No. 72, ASCE, 2nd edition, New York.

ASCE (1991). *Guidelines for electrical transmission line structural loading.* ASCE Manual 74, New York.

ASTM (1989). *Standard specification for steel wire strands.* ASTM Standard A475.

ASTM (1992). *Standard specification for zinc coated parallel and helical steel wire structural strands.* ASTM Standard A586–92.

Behncke, R. H., Milford, R. V., and White, H. B. (1994). ''High intensity wind and relative reliability based loads for transmission line design.'' Paper 22-205, CIGRE, Paris, August.

Bresler, B., Lin, T. Y., and Scalzi, J. B. (1960). *Design of steel structures.* 2nd edition, John Wiley, New York.

CAN/CSA (1986). *Antennas, towers and antenna supporting structures.* CSA Standard S37.M86, Canadian Structural Association, Rexdale, Ontario.

CAN/CSA (1992). *Zinc coated steel wire strands.* CAN/CSA Standard G12-1992, Canadian Structural Association, Rexdale, Ontario.

Gere, J. M. and Carter, W. O. (1962). ''Critical buckling loads for tapered columns.'' *J. Structural Division,* ASCE, 88 (ST1), February, 1–12.

IEEE (1985). *Trial-use guide for transmission structure foundation design.* IEEE Standard 691, New York (under revision).

IEEE   (1988).   *Guide to the assembly and erection of metal transmission structures.* IEEE Standard 951, New York.

IEEE   (1991).   *IEEE trial use design guide for wood transmission structures.* IEEE Standard 751, New York.

IEEE   (1991).   *Guide to the installation of foundations for transmission line structures.* IEEE Standard 977, New York.

Johnson, B. G.   (1976).   *Guide to stability design criteria for metal structures.* Column Research Council, 3rd edition, John Wiley, New York.

Kravitz, R. A. and Samuelson, A. J.   (1969).   "Tower designs for American Electric Power 765 kV Project." *J. of Power Division, ASCE* 95(2), October, 305–319.

NESC   (1993).   *National electrical safety code.* ANSI C2-1992, IEEE, New York.

Peabody, A. B. and Wekezer, J. W.   (1994).   "Buckling strength of wood power poles using finite elements." *J. of Structural Engrg. ASCE,* 120(6), 1893–1908.

Peyrot, A. H. and Goulois, A. M.   (1979).   "Analysis of cable structures." *J. of Computers and Structures,* 10, 805–813.

REA   (1992).   *Design manual for high voltage transmission lines.* REA Bulletin 1724E-200, United States Department of Agriculture, Washington, DC.

Salmon, C. G. and Johnson, J. E.   (1990).   *Steel structures, design and behavior,* Harper Collins, New York.

White, H. B.   (1993).   "Guyed structures for transmission lines." *Engng. Structures,* 15(4), 289–302.

Appendix B

NOTATION

A = guy cross-sectional area (mm^2 or in.^{2})

$ALPHA$ = guy slope—angle measured from horizontal to chord (degree)

CF = factor for effect of shear on buckling strength of laced mast

D = guy outside diameter (mm or in.)

d_a = pole top diameter (mm or in.)

d_g = pole ground diameter (mm or in.)

E = guy effective modulus of elasticity (MPa or ksi)

EHS = extra high strength grade

E_C = concrete material modulus of elasticity (MPa or ksi)

E_M = mast material modulus of elasticity (MPa or ksi)

E_P = pole material modulus of elasticity (MPa or ksi)

E_S = steel material modulus of elasticity (MPa or ksi)

F_G = guy tension force (kN or kips)

$GAMMA$ — angle defined in Fig. 5-5 (degree)

h = structure height to lower guy (m or ft)

H = horizontal load (kN or kips)

HF_G = horizontal component of tension force in guy (kN or kips)

HF_P = horizontal force in pole (kN or kips)

I_a = moment of inertia at top of wood pole (mm^4 or in.^{4})

I_C = moment of inertia of concrete pole section (mm^4 or in.^{4})

I_g = moment of inertia at ground level for wood pole (mm^4 or in.^{4})

I_M = moment of inertia of mast cross-section (mm^4 or in.^{4})

I_P = moment of inertia of pole cross-section (mm^4 or in.^{4})

K = equivalent pin-end length or unbraced length factor

K_G = horizontal guy stiffness—horizontal guy reaction HF_G caused by a unit horizontal displacement Δ (kN/m or kips/in.)

K_P = horizontal pole stiffness (kN/m or kips/in.)

l = longitudinal projection of guy on horizontal plane (m or ft)

L = guy chord length—from anchor to structure attachment point (m or ft)

M = pole base moment (kN-m or ft-kips)

P = vertical load in pole or mast (kN or kips)

P^* = Gere and Carter correction factor for buckling load

P_{CR} = theoretical buckling load (kN or kips)

PHI = one-half of line angle (degree)

PF_G = pretension force—value of tension in unloaded structure expressed as a fraction of RBS (percent)

RBS = rated breaking strength or ultimate tension capacity (kN or kips)

t = transverse projection of guy on horizontal plane (m or ft)

VF_G = vertical component of tension force in guy (kN or kips)

VF_P = vertical force in pole (kN or kips)

W = guy bare weight per unit length (N/m or lbs/ft)

Δ = horizontal displacement of structure attachment point (mm or in.)

INDEX

Alignment 61
Allowable tensions 50
Alumoweld 15
Anchor rods, installing 24, 60-61
Anchorage 21-25, 51, 57
Angle poles 4, 5, 57, 67
Attachment points 31-37, 44

Ball and cup forgings 53
Bisector guyed pole 68
Bolt slippage 40
Broken wire 42
Buckling 42, 43, 53; capacity 5, 64-66, 68-69; load calculations 41, 67; shapes 39, 40; strength 38-41

Cable behavior 27-31
Canadian Grade 1300 strand 15, 17
Chainette. *See* Cross rope suspension
Clearances 48-49
Column method analysis 36-37, 64, 66, 68, 69-70
Combination loads 73-74
Compressed fittings 17
Compression terminals 19
Computer modeling 42-45
Concrete anchor 22
Concrete poles 6, 35, 40, 45
Connections 50, 51, 52-53
Construction considerations 55-56
Crane erection 57-58
Cross guys 7
Cross rope suspension (CRS) 9-10, 13, 59

Dead-end poles 57, 63-66
Dead-end structures 3
Deadman anchors 21-22
Delta towers 9, 10, 42, 59, 75, 76-77
Design considerations 55
Direct embedded poles 57

Eccentric guy connection 44
Elastic instability 43
Elasticity. *See* Modulus of elasticity

Electrical clearances 48
Equivalent beam model 40-41
Erection methods, poles 56
Erection tolerances 60-61
Extra High Strength (EHS) strands 15, 16

Failure analysis 38
Fittings 17-20
Fixed base 6, 32-36, 37
Flemish loop 18
Force distribution 5, 45
Foundations 21-25; movement 35-36, 57
Four guy patterns 37-38, 47, 49
Frost heave 47

Galvanized high-strength strandings 15
Geometrical properties 27-28
Geometrically nonlinear analysis 43
Gere and Carter method 39, 68
Grouted anchors 23-25
Guy arrangements 5
Guy forces 71
Guy properties 70
Guy restrictions 48
Guying configurations 4, 72

H-frames 7-8, 43, 57, 60; structural design 53-54
Haunches 12
Head guys 33
Helical wire terminals 17, 18, 19
Helicopter erection 58
Hinged base 31-32
Hinged mast 32, 33, 37
Hinged Y 13-14

Ice loads 47, 50
In-line guys 33, 34-35, 63-67
Inspection 61
Installation 49, 56-57, 58-60
Internally guyed portal 8, 11

Lateral loads 35, 45
Latticed guyed delta 9, 10

Latticed masts 4, 40-41; with single attachment point 31-36
Latticed rigid frames 54
Latticed structures 8-9; connections 52-53
Line designer 55
Linear computer analysis 42-43
Log anchors. *See* Deadman anchors
Longitudinal loads 48, 73

Maintenance 61
Manual methods of analysis 42, 66, 67
Mast base 5-6, 52
Mast rotation 42
Masted towers 9-14, 54, 57-60
Masts 4-6, 31; design for guyed V 74-76
Materials 15-16
Mechanical clearances 48-49
Mechanical properties 27-28
Modeling 44-45
Modulus of elasticity 28
Multi-guy levels 37
Multi-pole structures 7-8. *See also* H-frames
Multiple guy attachment points 36-37

Ninety-degree angle pole 66-67
Nonlinear computer analysis 43, 54, 65-67, 69, 70

P-delta effect 42
Pinned connections 52
Pole base 5-6
Poles 56-57, 60; behavior, 70-71; connections to 52; erection methods 56; structural design 53-54; with single guy attachment point 31
Portals 8, 9, 10, 11, 59
Poured sockets 19
Poured zinc sockets 20
Preformed grips 17
Preload 35-36
Pretension 35-36, 42, 49-50, 57, 60

Rated Breaking Strengths (RBS) 15, 16, 49, 50
Rigid frames 8-9, 57-60

Rigid latticed portal 8
Rigid Y 9

Screw anchors 22-23
Second order analysis 43
Single guy attachment points 31-36
Single poles 4-6
Slack behavior 30
Slack guys 38, 61
Slope 27, 28, 30, 50-51, 70
Spread type anchor 22
Spun cast round poles 6
Static cast square poles 6
Steel poles 6, 34-35, 67-71; tapered 40; response 71
Strain structures 3
Strand configuration 16
Stress distribution 8
Stress-load relationship 43
Structural design 53-54
Strut method analysis 36-37, 64, 70
Stub poles 6-7, 44
Swaged fittings 17

Tangent dead-end pole 66
Tangent poles 57
Tangent structures 49
Tangent suspension towers 59
Taut behavior 30
Taut guys 38
Temperature effect 28, 36
Tensioning devices 20
Tensions 37-38, 50, 61, 73-74
Tensions versus displacement 29, 30
Terrain 3, 55
Three-guy system 29
Torsional resistance 9
Transverse loads 32, 34, 35, 36, 73-74
Tubular delta 10
Tubular steel poles. *See* Steel poles
Turn back loop 18

U-bolt clips 18
Ultimate bond stress, anchors 24-25
Unbalanced longitudinal load 73

V tower 9, 10, 11-12, 59, 72-76;

design of mast 74-75; with haunches
    12
Vertical loads 57, 73-74

Weather 47, 56
Wedge sockets 18
Wedge-assisted helical wire terminals 19
Wind loads 28, 47, 50, 72-73
Wire rope connections, efficiency of 51
Wires, EHS grade 16
Wires, grade 1300 17
Wood H-frames 8, 54
Wood poles 34-35, 37, 39, 43, 49, 63-
    67; behavior 66, 67; naturally
    grown 6; structural design 53-54;
    tapered 34

X-braces 7